Problem Solver II™

Integrating Problem Solving
with Your Math Curriculum

Teacher Resource Book
Grade 1

Shirley Hoogeboom
Judy Goodnow

McGraw Hill | Wright Group

Acknowledgments

We wish to thank Marj Santos for reviewing the manuscript and guiding the classroom testing.

Judy Goodnow has authored and coauthored over 100 books and software programs for mathematics and problem solving. She has taught children from kindergarten through sixth grade. She holds a bachelor of arts degree from Wellesley College, a master's degree from Stanford University, and a California Teaching Credential from San Jose State University.

Shirley Hoogeboom has authored and coauthored over 100 books for mathematics and language arts. She has been a classroom teacher, and has conducted workshops for teachers on problem solving and on using math manipulatives. She holds a bachelor of arts degree in Education from Calvin College, where she earned a Michigan Teaching Credential. She completed further studies at California State University, Hayward, where she earned a California Teaching Credential.

 **Wright Group**

Problem Solver II: Integrating Problem Solving with Your Math Curriculum, Teacher Resource Book, Grade 1
Copyright ©2004 Wright Group/McGraw-Hill
Text by Judy Goodnow and Shirley Hoogeboom
Illustrations by John Haslam
Design and Production by O'Connor Design

Problem Solver™ is a trademark of the McGraw-Hill Companies, Inc.

Wright Group/McGraw-Hill
One Prudential Plaza
Chicago, Illinois 60601
www.WrightGroup.com
Customer Service: 800-624-0822

Printed in the United States of America.

10 9 8 7 6 5 4 3 2 1

ISBN: 0-322-08805-4

The McGraw-Hill Companies

Contents

All About *Problem Solver II*

What's new about this program?

For years, teachers have been using *The Problem Solver* program to teach their students an organized approach to problem solving using ten strategies. These teachers began asking for more problems that they could tie in to their regular math textbook, thus integrating problem solving with all the content strands. *Problem Solver II* responds to those requests. It gives you hundreds of new problems that can be used with virtually any mathematics curriculum. It offers more detailed teaching plans, often showing students how to combine strategies to solve problems. New Student Workbooks give the children a place to record their work and their thinking. Because the instruction and the problems are organized by math strand as well as by problem-solving strategy, it is now easier than ever for you to coordinate problem solving with your daily mathematics program.

How do these materials teach problem solving?

Like the original program, *Problem Solver II* guides students through step-by-step instruction in mathematical problem solving. With these materials, you teach students the easy-to-learn Four-Step Method of problem solving (explained on page xii). You also model the use of ten basic problem-solving strategies (described on pages xiii–xvii), strategies that are useful at every grade level. The goal is to help students become competent, confident, and creative problem solvers. You can use *Problem Solver II* as an introduction to problem solving, as a review of the problem-solving process and strategies, or as a means of enriching students' experience with nonroutine problems.

Does this program support the NCTM Standards?

The standards articulated by the National Council of Teachers of Mathematics (NCTM) have been a guiding force in the development of *Problem Solver II*. In its *Principles and Standards for School Mathematics*, NCTM emphasizes the importance of problem solving for all students from pre-kindergarten through the twelfth grade. The problem-solving standard specifies four key goals for instructional programs: that they "enable all students to

- build new mathematical knowledge through problem solving;
- solve problems that arise in mathematics and in other contexts;
- apply and adapt a variety of appropriate strategies to solve problems;
- monitor and reflect on the process of mathematics problem solving." (NCTM 2000, 52)

The problems in this program involve all five content areas of the mathematics curriculum: Number and Operations, Algebra, Geometry, Measurement, and Data Analysis and Probability. *Problem Solver II* also incorporates the five process standards identified by NCTM: Problem Solving, Reasoning and Proof, Communication, Connections, and Representation. As students work on the problems, they learn to reason and to communicate in many different ways. They represent their work by writing, drawing pictures or diagrams, using objects, and making tables and charts. Students make connections among content areas as well as between mathematics and real-life situations.

How is this program organized?

The *Problem Solver II* program includes a Teacher Resource Book and Student Workbooks for each grade from 1 through 6.

The Teacher Resource Book begins with a series of 38 Teaching Problems that you use to introduce the ten strategies, either alone or in combination with a strategy that was taught previously. The Teaching Problems are carefully sequenced by both math content strand and level of difficulty. They are presented in pairs, giving students an opportunity to use the same strategy or combination of strategies twice. The Student Workbook parallels this lesson sequence, presenting all 38 Teaching Problems on two-page layouts.

Following the Teaching Problems in the Teacher Resource Book, you will find a bank of reproducible Practice Problems that offer additional opportunities for students to apply their problem-solving skills.

Do I need anything besides the books?

Common manipulatives are recommended for some Teaching Problems, as indicated by icons at the beginning of the lesson. For grade 1, these tools include cubes, Pattern Blocks, and play money. With a class of 24 students working in pairs, we suggest the following quantities:

Linker Cubes: 130 cubes each of ten colors

Pattern Blocks: 100 hexagons, 100 squares, and 200 each of the other blocks

Play money: 240 pennies, 192 nickels, 160 dimes, 128 quarters, and 32 half dollars

You may also want to have calculators available for student use, especially for Practice Problems that require lengthy calculations. Calculator use is always optional.

How can I link the materials with my math curriculum?

If you intend to coordinate the material in *Problem Solver II* with the lessons in your math textbook, the charts on the following pages can facilitate your planning. In particular, the chart of Math Content Strands and Skills highlights the specific mathematics skills embedded in each problem. To select the problems that best coordinate with your textbook units, you can scan the chart to find problems that involve the mathematics your students are currently studying.

Overview of Strategies and Skills in *Problem Solver II*, Grade 1

TEN PROBLEM-SOLVING STRATEGIES

Problem	Act Out or Use Objects	Use or Make a Picture or Diagram	Use or Make a Table	Make an Organized List	Guess and Check	Use or Look for a Pattern	Work Backwards	Use Logical Reasoning	Make It Simpler	Brainstorm
TEACHING PROBLEMS										
1	X									
2	X									
3			X							
4			X							
5	X					X				
6	X					X				
7	X				X					
8	X				X					
9	X					X				
10	X					X				
11	X							X		
12	X							X		
13	X			X						
14	X			X						
15	X			X					X	
16	X			X					X	
17	X	X								
18	X	X								
19	X		X							
20	X		X							
21		X								
22		X								
23	X							X		
24	X							X		
25	X						X			
26	X						X			
27		X					X			
28		X					X			
29			X			X				
30			X			X				
31	X							X		
32	X							X		
33		X						X		
34		X						X		
35										X
36										X
37	X							X		
38	X							X		
PRACTICE PROBLEMS										
39	X									
40	X									
41	X									
42			X							
43		X								
44		X								
45	X					X				
46	X					X				
47	X					X				

The strategies indicated here for the Practice Problems (39–100) are those which were used for solving similar Teaching Problems. However, the students' choice of strategy may vary from these.

MATH CONTENT STRANDS AND SKILL FOCUS

	Number and Operations	Algebra	Geometry	Measurement	Data Analysis and Probability	Skill Focus
TEACHING PROBLEMS						
1	■	■				Identify part/whole (how many in all); add; subtract
2	■	■				Identify part/whole (how many in all); add; subtract
3	■	■				Count by 2s; Identify part/whole (how many in all); add
4	■	■				Count by 3s; Identify part/whole (how many in all); add
5	■	■				Use patterns; count on; add
6	■	■				Use patterns; count on by 2s; add
7	■	■				Compare numbers (more than); identify part/whole; add; subtract
8	■	■				Compare numbers (more than); identify part/whole; add; subtract
9	■	■				Identify and use patterns; count on; add
10	■	■				Identify and use patterns; count on; add
11		■				Analyze data; identify order (last, behind, next to)
12		■				Analyze data; identify order (between, first)
13	■					Make different combinations of two coins
14	■					Make different combinations of four colors
15	■	■				Make different combinations of 5
16	■	■				Make different combinations of 6
17	■	■	■			Use number line; count forward and backwards
18	■	■	■			Use number line; count forward and backwards
19	■					Compare amounts (more/less); add coin values
20	■					Compare amounts (more/less); add coin values
21			■	■		Follow paths on a coordinate grid
22			■	■		Follow paths on a coordinate grid
23		■			■	Read picture graph; compare amounts (more than, the same)
24		■			■	Read bar graph; compare amounts (more/fewer); even numbers
25	■					Compare numbers (more than); add
26	■					Compare numbers (more than); add
27				■		Count hours on the clock; add; subtract
28				■		Count hours on the clock; add; subtract
29	■	■				Identify and use patterns; count by 4s
30	■	■				Identify and use patterns; count back by 2s
31			■		■	Compare geometric shapes (number/length of sides, square corners)
32			■		■	Compare geometric shapes (number/length of sides, square corners)
33	■				■	Use Venn diagram; sort data; add
34	■				■	Use Venn diagram; sort data; add; subtract
35					■	Analyze data; use spatial reasoning; add coin values
36					■	Analyze data; consider things that come in pairs
37	■					Identify equivalent costs; add
38	■					Identify equivalent costs; add
PRACTICE PROBLEMS						
39	■	■				Identify part/whole (how many in all); add; subtract
40	■	■				Identify part/whole (how many in all); add; subtract
41	■	■				Identify part/whole (how many in all); add; subtract
42	■	■				Count by 4s; identify part/whole (how many in all); add
43	■	■				Count by 5s; identify part/whole (how many in all); add
44	■	■				Count by 6s; identify part/whole (how many in all); add
45	■	■				Use patterns; count on by 2s; add
46	■	■				Use patterns; count on by 3s; add
47	■	■				Use patterns; count on by 2s; add

PRACTICE PROBLEMS

	Act Out or Use Objects	Use or Make a Picture or Diagram	Use or Make a Table	Make an Organized List	Guess and Check	Use or Look for a Pattern	Work Backwards	Use Logical Reasoning	Make It Simpler	Brainstorm
48	▨				▨					
49	▨				▨					
50	▨				▨					
51			▨			▨				
52			▨			▨				
53			▨			▨				
54	▨							▨		
55	▨							▨		
56	▨							▨		
57	▨			▨						
58	▨			▨						
59	▨			▨						
60	▨			▨					▨	
61	▨			▨					▨	
62	▨			▨					▨	
63		▨								
64		▨								
65		▨								
66	▨	▨	▨							
67	▨	▨	▨							
68	▨	▨	▨							
69		▨								
70		▨								
71		▨								
72		▨						▨		
73		▨						▨		
74		▨						▨		
75	▨						▨			
76	▨						▨			
77	▨						▨			
78		▨					▨			
79		▨					▨			
80		▨					▨			
81			▨			▨				
82			▨			▨				
83			▨			▨				
84	▨							▨		
85	▨							▨		
86	▨							▨		
87		▨						▨		
88		▨						▨		
89		▨						▨		
90										▨
91										▨
92										▨
93	▨							▨		
94	▨							▨		
95	▨							▨		
96				▨				▨		
97			▨			▨				
98		▨						▨		
99		▨						▨		
100			▨			▨				

MATH CONTENT STRANDS AND SKILL FOCUS

PRACTICE PROBLEMS

	Number and Operations	Algebra	Geometry	Measurement	Data Analysis and Probability	Skill Focus
48	■					Compare numbers (more than); identify part/whole; add; subtract
49	■					Compare numbers (more than); identify part/whole; add; subtract
50	■					Compare numbers (fewer than); identify part/whole; add; subtract
51	■	■				Identify and use patterns; count by 3s; add
52	■	■				Identify and use patterns; count on by 2s; add
53	■	■				Identify and use patterns; count back by 3s; subtract
54		■				Analyze data; identify order (first/last, the same as, between)
55		■				Analyze data; identify order (last, behind)
56		■				Analyze data; identify order (top/bottom, between)
57	■				■	Make different combinations of coins
58	■				■	Make different combinations of colors
59	■				■	Make different combinations of colors
60	■				■	Make different combinations of 6
61	■				■	Make different combinations of 7
62	■				■	Make different combinations of 8
63	■	■	■			Use a number line; count forward and backwards
64	■	■	■			Use a number line; count forward and backwards
65	■	■	■			Use a number line; count forward and backwards
66	■					Compare amounts (more/less); add coin values
67	■					Compare amounts (more/less); add coin values
68	■					Compare amounts (more/less); add coin values
69			■			Follow paths on a coordinate grid
70			■			Follow paths on a coordinate grid
71			■			Follow paths on a coordinate grid
72		■			■	Read picture graph; compare amounts (more than, same as)
73					■	Read bar graph; compare amounts (more/less); even numbers
74					■	Read bar graph; compare amounts (fewest, same as, more than)
75	■					Compare amounts (more than, same as); add
76	■					Compare amounts (more than); add
77	■					Compare amounts (more/fewer); add; subtract
78				■		Count hours on the clock; add; subtract
79				■		Count hours on the clock; add; subtract
80				■		Count hours on the clock; add; subtract
81	■	■				Identify and use patterns; add by 3s
82	■	■		■		Identify and use patterns; count hours on the clock; subtract by 5s
83	■	■				Identify and use patterns; count by 6s
84			■		■	Identify and compare geometric shapes (corner size)
85			■		■	Compare geometric shapes (number/length of sides, square corners)
86			■		■	Compare geometric shapes (number of sides, corner size)
87	■				■	Use Venn diagram; sort data; add; subtract
88	■				■	Use Venn diagram; sort data; add; subtract
89	■				■	Use Venn diagram; sort data; add; subtract
90			■		■	Use spatial reasoning; create specified geometric shape
91					■	Analyze data; consider things in groups of 2 and 4
92			■			Use spatial reasoning; create specified geometric shape
93	■					Identify equivalent costs; add
94	■					Identify equivalent costs; add
95	■					Identify equivalent costs; add
96	■					Identify part/whole; compare amounts (more, same as); add coin values
97	■	■				Identify and use patterns; count back by 3s; subtract
98	■				■	Use Venn diagram; sort data; add; subtract
99	■				■	Read bar graph; compare amounts (fewer, twice); identify odd number
100	■	■				Identify and use patterns; subtract by 4s

Teaching Suggestions

How should I present the lessons?

Lessons in *Problem Solver II* follow an easy-to-use format, with a two-page lesson plan for each Teaching Problem. If you are using these problems to introduce the problem-solving strategies to your students, you can simply use the pages in order. If your students are already familiar with the strategies, you might instead select problems related to the math skills they are currently studying.

The lesson plans offer a script to help you guide students through the problem-solving process, modeling the Four-Step Method and showing why a particular strategy (or combination of strategies) is useful. The script includes questions you can ask and shows sample student responses in italics; likely you will want to add further questions of your own as you hear how your students are responding. You can present the lessons to the entire class or to small groups; some teachers like to work at an overhead projector.

Before starting on the Teaching Problems, you may want to introduce separately the Four-Step Method that students will be using throughout the program. For a discussion of this method, see page xii.

How should I use the Student Workbooks?

Have the students work in their Student Workbooks while you present each lesson. To start the lesson, first read the problem aloud to the group, then have students read it again. In grade 1, the step-by-step questions in the Teacher Resource Book are not shown in the Student Workbook. However, as you present these questions orally to guide students through the problem-solving process, they will record their thinking on the workbook pages. In this way, their Student Workbook becomes a valuable problem-solving reference they can use as they work on the Practice Problems.

What if my students have different ways to solve the problems?

Students will likely offer a wide range of responses to the step-by-step questions. There are often many different ways to approach a given problem, and you should encourage students to share their thoughts. As the NCTM has stated:

> Sharing gives students opportunities to hear new ideas and compare them with their own and to justify their thinking. As students struggle with problems, seeing a variety of successful solutions improves their chance of learning useful strategies and allows them to determine if some strategies are more flexible and efficient [than others]. (NCTM 2000, 118)

The students will have even more freedom to think through their own approaches when they work on the Practice Problems.

How do the Practice Problems fit into the program?

Each lesson plan identifies three or four Practice Problems that are similar to the Teaching Problem students have just solved. You can reproduce any or all of these for follow-up work. Keep in mind, however, that no particular strategies are indicated on the Practice Problem page. This gives students the opportunity to analyze the problem themselves and choose the strategies that they believe will be most helpful for solving it.

Most of the Practice Problems give students an outline of the Four-Step Method: FIND OUT, CHOOSE STRATEGIES, SOLVE IT, and LOOK BACK. However, a series of problems at the end of the book is presented with no such guidance, giving students a chance to call up all their problem-solving skills—much as they will have to do in textbooks, on tests, and in daily life.

Several additional follow-up options are built into the
Problem Solver II program.

- Extension Problems offer further work that is related
  in some way to the Teaching Problem. Students solve
  these problems independently, then meet with a partner
  or in small groups to discuss their solution methods.
  This encourages them to share their different ways of
  thinking about and solving the problem.

- Having students write a problem that is similar to the
  one they just solved increases their understanding of
  problem-solving situations and solution methods.
  Students can exchange their problems for more practice.
  You might consider adding some of the student-created
  problems to the bank of Practice Problems.

- Some teachers like to give students one or more of the
  Writing Questions (page xx), asking them to tell how
  they solved the problem and to explain their thinking
  process in writing. These questions can also be used
  with the Practice Problems as a tool for assessment,
  as described on page xviii.

The Four-Step Method

The Four-Step Method is a systematic approach to problem solving that can be used for solving any problem. You may want to discuss these steps with your students before beginning work with the Teaching Problems. Understanding the purpose of each step can help students experience greater success.

Step 1: FIND OUT

The first step for students in solving a problem is to make sure they know what the problem is about and what they are being asked to find. Encourage them to try explaining the problem in their own words; this helps them better understand the information. They should ask themselves:

- What is happening in the problem?
- What do I have to find out to solve the problem?
- Are there any words or ideas I don't understand?
- What information can I use?
- Am I missing any information that I need?

Step 2: CHOOSE STRATEGIES

After students have identified what they are looking for and they know what information they have, they can make a plan for solving the problem. Now is the time to choose the strategy or combination of strategies that they think will be most helpful. They will find that there is often more than one way to solve the problem. In some cases, the problem may have to be broken down into smaller problems before the larger problem can be solved.

Step 3: SOLVE IT

Students now use the strategy or strategies they have chosen to solve the problem. It is very important that they record their work in a way that lets them see what they have completed. It is possible they will discover that the strategy they chose is not as helpful as they thought it would be. Emphasize that they should not be discouraged, but rather choose a different strategy and try again.

Step 4: LOOK BACK

After students have solved the problem, they should always check their answer by reading the problem again, looking back over each step, and checking their calculations. They should ask themselves:

- Did I answer the question asked in the problem?
- Is more than one answer possible?
- Is my math correct?
- Does my answer make sense? Is it reasonable?
- Can I explain why I think my answer is correct?

Ten Problem-Solving Strategies

The following ten strategies taught in *Problem Solver II* are useful for solving many kinds of problems. Students may discover other strategies on their own; when this happens, encourage them to share their discoveries with their classmates.

Act Out or Use Objects

Acting out a problem or having objects to move around gives students visual images of the problem and of the steps they must take to solve it. Using this strategy allows students to visualize arrangements, combinations, and relationships in the elements of a problem. Common manipulatives such as Pattern Blocks, cubes, play money, and even scraps of colored paper work well to represent numbers of items and colors.

Example Joy has 12 coins in all. She has 4 more nickels than pennies. How many nickels and pennies does Joy have?

Students can use play money to work out a solution. They might start with 12 nickels, then exchange one nickel at a time for a penny until they find the combination that gives Joy 4 more nickels than pennies.

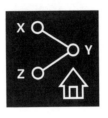

Use or Make a Picture or Diagram

Some problems give a picture, diagram, or map as part of the data. For other problems, it may be helpful for students to draw their own picture or diagram. A simple pictorial representation can often help students understand and work with the data in the problem.

Example Sue and Willy are playing a board game. Sue's game marker is on START. She moves her marker forward 5 spaces on the track. Next, she moves her marker back 3 spaces. Next, she moves her marker forward 8 spaces. Willy's game marker is on space 9. Who is ahead, Sue or Willy?

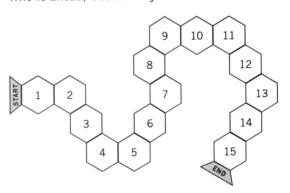

A diagram of the game track is helpful for solving this problem. Students can count forward and backwards to show Sue's moves.

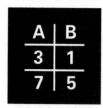

Use or Make a Table

In some problems, students may need to use data from a table or a chart. In other problems, they may need to keep track of data in an orderly way. Making their own tables by listing key numbers in sequence can help students find missing data and discover or extend number patterns. This strategy is often used in combination with other strategies.

Example Crystal and Holly made puppets. The girls glued 2 eyes on each puppet. They used 18 eyes in all. How many puppets did the girls make?

Number of Puppets	Number of Eyes
1	2
2	4
3	6
4	8
5	10
6	12
7	14
8	16
9	18

Creating a table for the problem data helps students see that when the girls have used 18 eyes, they will have glued eyes on 9 puppets.

Make an Organized List

An organized list is a systematic way of recording a series of computations or exploring combinations of items. This strategy helps organize a student's thinking. A list makes it easy to see what has been done and to identify steps that still need to be completed. An organized list is especially helpful when a student wants to consider all the possibilities in order to find those that fit the problem.

Example Tomas has 3 yellow balloons, 3 green balloons, and 2 blue balloons. He ties the balloons in groups of 2. Each group is different from every other group. What colors are the balloons in each group?

yellow, yellow

green, green

blue, blue

yellow, green

Making an organized list helps students keep track of different combinations of the 8 balloons, making sure that all of the balloons have been included and that no groups are exactly alike.

Guess and Check

Sometimes it is helpful to guess an answer when trying to solve a problem. When students use this strategy, they guess a number and then check to see if it fits with the other data given in the problem. If it does not, they decide whether their guess was too high or too low, then try to come closer with their next guess. They keep guessing and checking until they find a correct answer.

Example Jan and Jack are playing The Fishing Game. They catch 9 fish in all. The fish are red and green. They catch 3 more red fish than green fish. How many fish of each color do they catch?

First guess:

The students' first guess of 5 green fish means there would be 4 red fish. When they check the problem clues, they find that this cannot be right because there are not 3 more red fish than green fish. Their next guess for green fish needs to be smaller. Guessing and checking in this way will lead to the correct answer of 6 red fish and 3 green fish.

Use or Look for a Pattern

Looking for patterns is a very important strategy for problem solving; it is used to solve many different kinds of problems. A pattern occurs when a relationship is repeated again and again. A pattern may be numerical, visual, or behavioral. Identifying a pattern enables students to predict what will "come next" and what will happen again and again in the same way. In some problems, a pattern is given and students use it to solve the problem. In other problems, students must identify and extend the pattern in order to find a solution. Making a table often reveals patterns in data, so these two strategies are frequently used in combination.

Example On Monday the bears had 13 cookies in their cookie jar. On Tuesday there were 11 cookies in the jar. On Wednesday there were 9 cookies in the jar. Cookies keep disappearing each day in the same way. When will the cookie jar be empty?

Day	Number of Cookies
Monday	13
Tuesday	11
Wednesday	9
Thursday	7
Friday	5
Saturday	3
Sunday	1
Monday	0

Placing the information into a table will make it easier for students to see that the number of cookies decreases by 2 each day, and that the jar will be empty on Monday.

Work Backwards

To solve certain problems, it is necessary to begin with data presented at the end of the problem and then work backwards.

Example Carlos is sorting puzzle pieces. He has 4 more blue pieces than green pieces. He has 5 more green pieces than red pieces. He has 1 red piece. How many blue puzzle pieces does Carlos have?

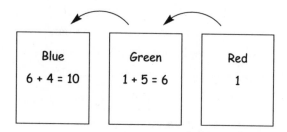

Blue	Green	Red
6 + 4 = 10	1 + 5 = 6	1

Starting with the last information given, 1 red piece, students would work backwards, using the clues to find how many puzzle pieces of each color Carlos has.

Use Logical Reasoning

Logical reasoning is actually involved in all problem solving. However, certain problems include or imply conditional statements, such as: "if something is true, then..." or "if something is not true, then...." Problems like these require more formal logical reasoning as students step their way through the clues.

Example Chico has 4 model race cars lined up. A blue car is between 2 red cars. The green car is not first in line. What color is each race car in the line?

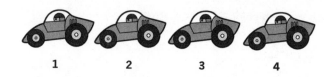

1   2   3   4

Students can step their way through this problem, using cubes to stand for the colored race cars. They can begin by placing a blue cube between 2 red cubes. Then they can reason, "If the green car is not first in line, it must be the last car in line." So they will reason that the cars are lined up in this order: red, blue, red, green.

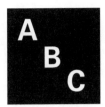

Make It Simpler

Sometimes a problem can be made simpler by reducing large numbers to small numbers, or by reducing the number of items given in a problem. Having a simpler representation can make it easier to recognize the operation or process that can be used to solve the more complex problem. The simpler representation may even reveal a pattern that can be used to solve the problem.

Example Reggie has 2 pockets in his pants. He puts 5 stones into his pockets. He puts at least 1 stone into each pocket. What are all the different ways that Reggie could put the stones into his pockets?

Pocket A	Pocket B
1	2
2	1

Pocket A   Pocket B

To make it simpler, students can try the problem with fewer stones, first putting just 3 stones into the 2 pockets. Once they see how to organize the information, they are better prepared to find all possible combinations of 5 stones in the 2 pockets.

Brainstorm

This strategy is often used when all else fails. When students cannot think of a similar problem they have solved before, and they cannot think of another strategy to use, brainstorming is a good strategy to try. It leads students to look at a problem in new and unusual ways. They should be encouraged to open up, stretch, allow for inspiration, be creative, be flexible, and keep on trying until a light goes on!

Example What could these codes mean?
10 P in a D   5 P in a N

parts   pies   paints

people in a _____

pieces in a _____

pennies in a _____ DIME!

Although this would not be considered a true mathematics problem, and the brainstorming involves language skills, it requires creative thinking about quantitative relationships to figure out that 10 P in a D stands for 10 pennies in a dime, and 5 P in a N stands for 5 pennies in a nickel.

Assessment

Before starting to use *Problem Solver II* with your students, you can assess their knowledge of the problem-solving process by giving them Practice Problems that involve different strategies. Assessing their ability to solve these test problems will help you decide which problems to use with your students. Such a pretest will also give you a baseline for evaluating student progress if you repeat this procedure at the end of the year.

As they learn the problem-solving process, students need to understand that the formulation of a plan and the execution of that plan are as vital as finding the correct answer. There are various ways to assess students' abilities in these areas.

- Observe students as they solve problems alone or with a partner or small group. You will gain meaningful insights into their ability to select and use appropriate techniques and their understanding of mathematical concepts. You will also be able to assess their ability to articulate, to listen, to be flexible, and to rethink.

- Have students write an answer to one or more Writing Questions (page xx) after they solve a Practice Problem. You might duplicate the Writing Questions on the back of a Practice Problem and indicate which ones you would like the student to write about.

- Collect student work on the Practice Problems and use a rubric to evaluate what they have done. This "holistic" assessment tool helps you evaluate the quality of students' work as a whole, not just the correctness of their answers. Two sample types of rubric are shown on the following page, or your school may have its own.

Whatever approach you take, make sure that students understand how their work is being assessed and involve them in the assessment process. Share your assessment form with them so they know what is expected. On at least some occasions, have students assess themselves. For this purpose, you might simplify your rubric, or have them use a simple 1–5 rating scale such as this:

1 – I don't know what to do.
2 – I'm beginning to understand.
3 – I'm doing OK, but I'm not sure about every step.
4 – I get it, and I'm sure my answer is right.
5 – I get it, and I gave a great explanation!

4-Point Rubric

You could use this rubric to evaluate student's work on the problem as a whole, rating it from 1 to 4.

1: Needs Improvement
- Uses inappropriate strategies
- Makes major errors
- Shows inadequate explanation of thinking
- Presents incorrect solution

2: Fair
- Gives adequate response
- Makes some errors
- Gives incoherent or unclear explanation
- Shows partial understanding of mathematical ideas
- May use appropriate strategy
- Shows some parts of solution process
- May have incorrect solution

3: Good
- Uses strategy or strategies appropriately
- Gives complete response
- Gives good clear explanation of thinking
- Shows understanding of mathematical concepts
- Shows solution process, including appropriate graphs, diagrams, etc.
- Presents correct solution

4: Outstanding
- Same as Good plus the following:
- Shows original and creative thinking
- Gives strong explanation of reasoning
- May use multiple strategies or multiple approaches
- Demonstrates outstanding grasp of mathematical ideas
- Exceeds expectations

Analytic Problem-Solving Rubric

With this rubric, you could evaluate each part of the problem-solving process separately, with points scored as shown.

Understands problem
- **0** Shows no understanding
- **1** Shows partial understanding
- **2** Shows full understanding, extracts relevant information, and understands question

Chooses strategies
- **0** Chooses inappropriate strategies or no strategies
- **1** Chooses appropriate strategies, but does not use them effectively
- **2** Chooses one or more appropriate strategies and uses them effectively

Solves the problem
- **0** Does not formulate a plan
- **1** Has a plan, but shows errors in thinking
- **2** Has a thoughtful plan and executes it well

Finds answer
- **0** Has no answer
- **1** Has an answer, but there are errors
- **2** Has correct answer

Looks back
- **0** Has inadequate or no explanation
- **1** Has explanation that is not clear, but demonstrates some understanding
- **2** Has explanation that shows clear understanding of mathematical ideas

Tell about or explain the problem in your own words.

What strategy did you choose? Why?

How did you solve the problem? Tell about your thinking.

Why does your answer make sense?

Thinking Questions
Questions to think about as you are solving problems

FIND OUT

What is happening in the problem?

What do I have to find out to solve the problem?

Are there any words or ideas I don't understand?

What information can I use?

Am I missing any information that I need?

CHOOSE STRATEGIES

Have I solved a problem like this before?

What strategies helped me solve it?

Can I use the same strategies for this problem?

SOLVE IT

What information should I start with?

Do I need to add or subtract?

How can I organize the information that I use or find?

Is the strategy I chose helpful?

Would another strategy be better?

Do I need to use more than one strategy?

Is my work easy to read and understand? Is it complete?

LOOK BACK

Did I answer the question that was asked in the problem?

Is more than one answer possible?

Is my math correct?

Does my answer make sense?

Can I explain why I think my answer is correct?

Act Out or Use Objects

 Each pair needs: 10 cubes (any color)

1 The Fur and Fuzz toy store has 8 toy cats. The cats are on 3 tables. There are 5 cats in all on table A and table B. There are 7 cats in all on table B and table C. How many cats are on each table?

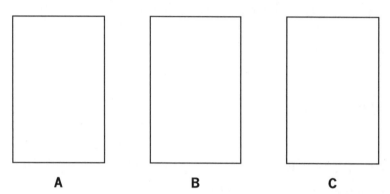

A   B   C

FIND OUT

- **What is the problem about?** Encourage students to restate the problem in their own words.

- **What do you have to find out to solve the problem?** *How many cats are on each table*

- **Find out what the problem tells you.**

 How many cats does the toy store have? *8 cats*

 How many tables are the cats on? *3 tables*

 How many cats in all are on table A and table B? *5 cats*

 How many cats in all are on table B and table C? *7 cats*

CHOOSE STRATEGIES

You can *Act Out or Use Objects*. Use cubes to show the cats.

SOLVE IT

Put cubes on the tables to show the cats.

1. **How many cubes do you need?** *8 cubes*

2. **Put some cubes on each table.** Answers to the following questions will vary, depending on how many cubes the students put on each table. The sample answers are based on putting 4 cubes on table A, 2 cubes on table B, and 3 cubes on table C.

 How many cubes did you put on table A? *4 cubes* **Table B?** *2 cubes* **Table C?** *3 cubes*

 How can you find the number of cubes in all on table A and table B? *Add them together.*

 TEACHING TIP

Be sure that students leave the cubes on each table and count them in place when they add together the numbers of cubes on two tables.

3. **Count all the cubes on table A and table B.
   Do they show 5 cats in all?** *No, they show 6 cats.*

4. **Count all the cubes on table B and table C.
   Do they show 7 cats in all?** *No, they show 5 cats.*

5. **If your answers are not right, change the cubes on
   the tables. Do this until you get the right numbers.**

6. **How many cats are on each table?**

   Solution: *Table A – 1 cat, table B – 4 cats,
   table C – 3 cats*

LOOK BACK

Students should read or listen to the problem
again and check their work. Encourage them to
ask themselves, **Did I answer the question that
was asked? Is my answer right?**

EXTENSION PROBLEM

**The Fur and Fuzz toy store has 10 toy bears.
The bears are in 3 baskets. There are 5 bears in
all in basket A and basket B. There are 8 bears
in all in basket B and basket C. How many bears
are in each basket?**

Solution: *Basket A – 2 bears, basket B – 3 bears,
basket C – 5 bears*

TALK ABOUT IT

Have students talk with a partner or with a
group about how they solved the Extension
Problem. Students can share their different ways
of thinking. Ask a question like, **How did using
cubes help you find the answer?**

WRITE YOUR OWN PROBLEM

Have students write similar problems of their
own. Students can exchange problems and
solve them.

PRACTICE

Similar Practice Problems: 39, 40, 41

When you give students a Practice Problem, ask
questions such as, **Have you solved a problem like
this before? What strategies helped you solve it?**

Act Out or Use Objects

 **Each pair needs: 14 cubes
(any colors)**

2 Alexa has 12 chocolate chip cookies. She puts them on 3 plates. Now there are 8 cookies in all on the first plate and second plate. There are 6 cookies in all on the second plate and third plate. How many cookies are on each plate?

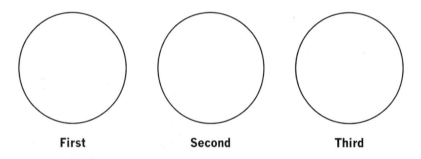

First   Second   Third

FIND OUT

- **What is the problem about?** Encourage students to restate the problem in their own words.

- **What do you have to find out to solve the problem?** *How many cookies are on each plate*

- **Find out what the problem tells you.**

  How many cookies does Alexa have? *12 cookies*

  How many plates are there? *3 plates*

  How many cookies are there in all on the first plate and second plate? *8 cookies*

  How many cookies are there in all on the second plate and third plate? *6 cookies*

CHOOSE STRATEGIES

You can *Act Out or Use Objects*. Use cubes to show the cookies.

SOLVE IT

Put cubes on the plates to show the cookies.

1. **How many cubes do you need?** *12 cubes*

2. **Put some cubes on each plate.** Answers to the following questions will vary, depending on how many cubes the students put on each plate. The sample answers are based on putting 5 cubes on the first plate, 3 cubes on the second plate, and 6 cubes on the third plate.

   How many cubes did you put on the first plate? *5 cubes* **Second plate?** *3 cubes* **Third plate?** *6 cubes*

   How can you find the number of cubes in all on the first plate and second plate? *Add them together.*

3. **Count all the cubes on the first plate and second plate. Do they show 8 cookies in all?** *Yes.*

🍎 TEACHING TIP

Be sure that students leave the cubes on each plate and count them in place when they add together the numbers of cubes on two plates.

4. **Count all the cubes on the second plate and third plate. Do they show 6 cookies in all?**
   No, they show 9 cookies.

5. **If your answers are not right, change the cubes on the plates. Do this until you get the right numbers.**

6. **How many cookies are on each plate?**

   Solution: *First plate – 6 cookies, second plate – 2 cookies, third plate – 4 cookies*

LOOK BACK

Students should read or listen to the problem again and check their work. Encourage them to ask themselves, **Did I answer the question that was asked? Is my answer right?**

EXTENSION PROBLEM

Alexa has 13 blueberry muffins. She puts them on 3 trays. Now there are 12 muffins in all on the first tray and second tray. There are 9 muffins in all on the second tray and third tray. How many muffins are on each tray?

Solution: *First tray – 4 muffins, second tray – 8 muffins, third tray – 1 muffin*

TALK ABOUT IT

Have students talk with a partner or with a group about how they solved the Extension Problem. Students can share their different ways of thinking. Ask a question like, **How did using the cubes help you find the answer?**

WRITE YOUR OWN PROBLEM

Have students write similar problems of their own. Students can exchange problems and solve them.

PRACTICE

Similar Practice Problems: 39, 40, 41

When you give students a Practice Problem, ask questions such as, **Have you solved a problem like this before? What strategies helped you solve it?**

Use or Make a Table

3 Crystal and Holly made puppets. The girls glued 2 eyes on each puppet. They used 18 eyes in all. How many puppets did the girls make?

FIND OUT

- **What is the problem about?** Encourage students to restate the problem in their own words.

- **What do you have to find out to solve the problem?** *How many puppets the girls made*

- **Find out what the problem tells you.**

  How many eyes did the girls glue on each puppet? *2 eyes*

  How many eyes in all did the girls use? *18 eyes*

CHOOSE STRATEGIES

You can *Use or Make a Table*. The table will help you keep track of the eyes on the puppets the girls made.

🍎 **TEACHING TIP**

Help children prepare to solve this problem by having them count by 2s up to 20. To prepare for the Extension Problem, have them count by 3s up to 24.

SOLVE IT

Look at the table that has been started.

Number of Puppets	Number of Eyes
1	2
2	4
3	6
4	8
5	10
6	12
7	14
8	16
9	18

1. **What will you keep track of in the first column of the table?** *The number of puppets the girls made*

   What will you keep track of in the second column? *The number of eyes on those puppets*

2. **How many eyes are there on 1 puppet?** *2 eyes* **Find that number in the table.**

3. **How many eyes are there on 2 puppets?** *4 eyes*
Write that number in the table.

   Help students understand that they can continue filling in the table in the same way. They need to add 2 eyes for each additional puppet, and they will record the new total in each row.

4. **How many eyes are there on 3 puppets?** *6 eyes*
Write that number in the table.

5. **Finish the table. Keep writing numbers until 18 eyes have been used.**

6. **How many puppets did the girls make?**

   Solution: *9 puppets*

LOOK BACK

Students should read or listen to the problem again and check their work. Encourage them to ask themselves, **Did I answer the question that was asked in the problem? Is my answer right?**

EXTENSION PROBLEM

Crystal and Holly are sewing 3 buttons on each puppet. When they have sewed 21 buttons on the puppets, they stop for lunch. How many puppets have buttons on them?

Solution: *7 puppets*

TALK ABOUT IT

Have students talk with a partner or small group about how they solved the Extension Problem. Students can share their different ways of thinking. Ask questions like, **How did your table help you find the answer? Did you use anything else to help you find the answer?**

WRITE YOUR OWN PROBLEM

Have students write similar problems of their own. Students can then exchange problems and solve them.

PRACTICE

Similar Practice Problems: 42, 43, 44

When you give students a Practice Problem, ask questions such as, **Have you solved a problem like this before? What strategies helped you solve it?**

4 The Rolling Robots Race is next week. Saul and his friends are making robots. They put 3 wheels on each robot. They use 24 wheels in all. How many robots do the children make?

FIND OUT

- **What is the problem about?** Encourage students to restate the problem in their own words.

- **What do you have to find out to solve the problem?** *How many robots the children made*

- **Find out what the problem tells you.**

  How many wheels do Saul and his friends put on each robot? *3 wheels*

  How many wheels in all do the children put on robots? *24 wheels*

CHOOSE STRATEGIES

You can *Use or Make a Table*. The table will help you keep track of the wheels on the robots.

🍎 **TEACHING TIP**

Help children prepare to solve this problem by having them count by 3s up to 36.

SOLVE IT

Look at the table that has been started.

Number of Robots	Number of Wheels
1	3
2	6
3	9
4	12
5	15
6	18
7	21
8	24

1. **What will you keep track of in the first column of the table?** *The number of robots*

   What will you keep track of in the second column? *The number of wheels on those robots*

2. **How many wheels do the children put on 1 robot?** *3 wheels* **Find that number in the table.**

3. **How many wheels do they put on 2 robots?**
   6 wheels **Write that number in the table.**

   Help students understand that they can continue filling in the table in the same way. They need to add 3 wheels for each additional robot, and they will record the new total in each row.

4. **How many wheels do they put on 3 robots?**
   9 wheels **Write that number in the table.**

5. **Finish the table. Keep writing numbers until there are 24 wheels.**

6. **How many robots do the children make?**

   Solution: *8 robots*

LOOK BACK

Students should read or listen to the problem again and check their work. Encourage them to ask themselves, **Did I answer the question that was asked in the problem? Is my answer right?**

EXTENSION PROBLEM

Saul and his friends keep making more robots and putting 3 wheels on each one. When they finish, they have used 33 wheels in all. How many robots in all did the children make?

Solution: *11 robots*

TALK ABOUT IT

Have students talk with a partner or small group about how they solved the Extension Problem. Students can share their different ways of thinking. Ask questions like, **How did your table help you find the answer? Did you use anything else to help you find the answer?**

WRITE YOUR OWN PROBLEM

Encourage students to write similar problems of their own. Students can then exchange problems and solve them.

PRACTICE

Similar Practice Problems: 42, 43, 44

When you give students a Practice Problem, ask questions such as, **Have you solved a problem like this before? What strategies helped you solve it?**

Use or Look for a Pattern
Act Out or Use Objects

 Each pair needs:
16 nickels

5 Amy takes care of pets to earn money. On Monday she got 1 nickel.
On Tuesday she got 2 nickels. On Wednesday she got 3 nickels.

Monday

Tuesday

Wednesday

Each day Amy gets 1 more nickel than she got the day before.
How many nickels will she get on Friday?

FIND OUT

- **What is the problem about?** Encourage students
  to restate the problem in their own words.

- **What do you have to find out to solve the problem?**
  How many nickels Amy will get on Friday

- **Find out what the problem tells you.**

  How many nickels did Amy get on Monday?
  1 nickel

  How many nickels did she get on Tuesday?
  2 nickels

  How many nickels did she get on Wednesday?
  3 nickels

  **What do you know about how many nickels Amy
  gets each day?** *Each day she gets 1 more nickel
  than she got the day before.*

CHOOSE STRATEGIES

You can *Use or Look for a Pattern* and *Act Out or
Use Objects*. Use play money to show the nickels.
Use the pattern to help you solve the problem.

🍎 **TEACHING TIP**

Ask students to explain what a pattern is and to look for
patterns around them. Help them to understand that in
a pattern, something happens again and again in the same
way. Sometimes a pattern can be found in a sequence of
numbers that grow larger or smaller in the same way.

SOLVE IT

Use play money. Draw circles to show the nickels for each day.

Monday ◯

Tuesday ◯◯

Wednesday ◯◯◯

Thursday ◯◯◯◯

Friday ◯◯◯◯◯

1. **How many nickels did Amy get on Monday?**
   1 nickel

2. **How many nickels did she get on Tuesday?**
   2 nickels

3. **How many nickels did she get on Wednesday?**
   3 nickels

4. **How many more nickels does she get each day than she got the day before?** *1 more*

5. **How many nickels will Amy get on Thursday?**
   4 nickels

6. **How many nickels will she get on Friday?**
   Solution: *5 nickels*

LOOK BACK

Students should read or listen to the problem again and check their work. Encourage them to ask themselves, **Did I answer the question that was asked in the problem? Is my answer right?**

EXTENSION PROBLEM

Amy has 15 pennies. She gives her little brother 1 penny on Tuesday. She gives him 2 pennies on Wednesday. Each day she gives him 1 more penny than she gave him the day before. On what day will Amy give her last penny to her brother?

Solution: *Saturday*

TALK ABOUT IT

Have students talk with a partner or small group about how they solved the Extension Problem. Students can share their different ways of thinking. Ask a question like, **How did using the pattern and play money help you solve the problem?**

WRITE YOUR OWN PROBLEM

Have students write similar problems of their own. Students can then exchange problems and solve them.

PRACTICE

Similar Practice Problems: 45, 46, 47

When you give students a Practice Problem, ask questions such as, **Have you solved a problem like this before? What strategies helped you solve it?**

**Each pair needs: 30 cubes
(any colors)**

6 Ken is starting to collect baseball cards. The first week, he buys 1 card.
The second week, he buys 3 cards. The third week, he buys 5 cards.

First week ☐

Second week ☐ ☐ ☐

Third week ☐ ☐ ☐ ☐ ☐

**Each week Ken buys 2 more cards than he bought the week before.
How many cards will Ken buy in the fifth week?**

FIND OUT

- **What is the problem about?** Encourage students to restate the problem in their own words.

- **What do you have to find out to solve the problem?**
  How many cards Ken will buy in the fifth week

- **Find out what the problem tells you.**

  How many cards does Ken buy the first week?
  1 card

  How many cards does he buy the second week?
  3 cards

  How many cards does he buy the third week?
  5 cards

  What do you know about the way Ken keeps buying baseball cards? *Each week he buys 2 more cards than he bought the week before.*

CHOOSE STRATEGIES

You can *Use or Look for a Pattern* and *Act Out or Use Objects.* Use cubes to show the cards. Use the pattern to help you solve the problem.

🍎 **TEACHING TIP**

Ask students to use cubes to make a pattern. Help them to understand that in a pattern, something happens again and again in the same way. Sometimes a pattern can be found in a sequence of shapes or in numbers that grow larger or smaller in the same way.

SOLVE IT

Use cubes to show the cards. Draw boxes to show the cards he buys each week.

First week ☐

Second week ☐☐☐

Third week ☐☐☐☐☐

Fourth week ☐☐☐☐☐☐☐

Fifth week ☐☐☐☐☐☐☐☐☐

1. **How many cards does Ken buy the first week?**
   1 card

2. **How many cards does he buy the second week?**
   3 cards

3. **How many cards does he buy the third week?**
   5 cards

4. **How many more cards does Ken buy each week than he bought the week before?** *2 more*

5. **Keep using the pattern and showing cards. Stop when you know how many cards Ken will buy in the fifth week.**

6. **How many cards will Ken buy in the fifth week?**

   Solution: *9 cards*

LOOK BACK

Students should read or listen to the problem again and check their work. Encourage them to ask themselves, **Did I answer the question that was asked in the problem? Is my answer right?**

EXTENSION PROBLEM

Ken keeps buying baseball cards in the same way. How many cards will he buy in the seventh week?

Solution: *13 cards*

TALK ABOUT IT

Have students talk with a partner or small group about how they solved the Extension Problem. Students can share their different ways of thinking. Ask a question like, **How did using the pattern and the cubes help you solve the problem?**

WRITE YOUR OWN PROBLEM

Have students write similar problems of their own. Students can then exchange problems and solve them.

PRACTICE

Similar Practice Problems: 45, 46, 47

When you give students a Practice Problem, ask questions such as, **Have you solved a problem like this before? What strategies helped you solve it?**

Guess and Check
Act Out or Use Objects

 Each pair needs: 20 cubes
(10 red and 10 green)

7 Jan and Jack are playing The Fishing
Game. They catch 9 fish in all. The fish
are red and green. They catch 3 more
red fish than green fish. How many fish
of each color do they catch?

FIND OUT

- **What is the problem about?** Encourage students
to restate the problem in their own words.

- **What do you have to find out to solve the problem?**
How many fish of each color they catch

- **Find out what the problem tells you.**

  How many fish do they catch in all? *9 fish*

  What colors are the fish? *Red and green*

  **How many more red fish than green fish do they
  catch?** *3 more*

CHOOSE STRATEGIES

**You can *Guess and Check* and *Act Out or Use
Objects*. Use cubes to show the fish.**

SOLVE IT

Put cubes on the fish to show the colors.

Answers to the following questions will vary,
depending on whether the first guess is for green
or blue. The sample answers are based on a first
guess of 5 for green fish.

1. **Make a guess for the red fish or the green fish.
   What number is your guess?** 5 **What color is the
   fish?** *Green* **Put this number of cubes on the fish.**

🍎 TEACHING TIP

Encourage the students to talk about why they would choose
green or red for their starting guess. Help them to see that
they can start with either color. They know there are 3 more
red fish than green fish, so they can either add 3 to a guess
for green to find how many red, or subtract 3 from a guess
for red to find how many green.

2. **Then what is the number for the other color?**
   4 red **Put this number of cubes on the fish.**

3. **Count up your red and green cubes. Do your cubes show 3 more red fish than green fish?** *No.*

4. **If your answers are not right, guess again. Do this until you get the right numbers. Should your next guess be larger or smaller?** *Smaller*

5. **How many fish of each color do they catch?**

   Solution: *6 red fish, 3 green fish*

LOOK BACK

Students should read or listen to the problem again and check their work. Encourage them to ask themselves, **Did I answer the question that was asked? Is my answer right?**

EXTENSION PROBLEM

Annie and Hoshi play the same game. They catch 13 fish in all. They catch 5 fewer green fish than red fish. How many fish of each color do they catch?

Solution: *9 red fish, 4 green fish*

TALK ABOUT IT

Have students talk with a partner or with a group about how they solved the Extension Problem. Students can share their different ways of thinking. Ask a question like, **How did you use a wrong guess to help you make another guess?**

WRITE YOUR OWN PROBLEM

Have students write similar problems of their own. Students can exchange problems and solve them.

PRACTICE

Similar Practice Problems: 48, 49, 50, 96

When you give students a Practice Problem, ask questions such as, **Have you solved a problem like this before? What strategies helped you solve it?**

Guess and Check
Act Out or Use Objects

 Each pair needs:
15 nickels and 15 pennies

8 Joy has 12 coins in all. She has 4 more nickels than pennies. How many nickels and pennies does Joy have?

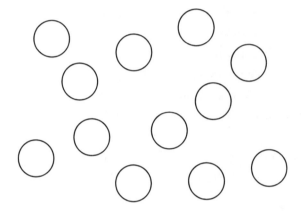

FIND OUT

- **What is the problem about?** Encourage students to restate the problem in their own words.

- **What do you have to find out to solve the problem?** *How many nickels and pennies Joy has*

- **Find out what the problem tells you.**

  How many coins does Joy have? *12 coins*

  What kinds of coins does she have? *Nickels and pennies*

  How many more nickels than pennies does she have? *4 more nickels*

CHOOSE STRATEGIES

You can *Guess and Check* and *Act Out or Use Objects*. Use play money to show the coins.

SOLVE IT

Put play money on the circles to show the coins.

Answers to the following questions will vary, depending on whether the first guess is for nickels or pennies. The sample answers are based on a first guess of 8 pennies.

1. **Make a guess for the nickels or the pennies. What number is your guess?** 8 **Which coins?** *Pennies* **Put this number of coins on the circles.**

 TEACHING TIP

Encourage the students to talk about which coin they would choose for their starting guess, nickels or pennies. Help them to see that they can start with either coin.

2. **Then what is the number for the other coins?**
   4 nickels **Put these coins on the circles.**

3. **Are there more nickels than pennies?** *No.*

4. **If your answer is not right, guess again. Do this until you get the right number of coins. Should your next guess be larger or smaller?** *Smaller*

5. **How many nickels and pennies does Joy have?**

   Solution: *8 nickels, 4 pennies*

LOOK BACK

Students should read or listen to the problem again and check their work. Encourage them to ask themselves, **Did I answer the question that was asked? Is my answer right?**

EXTENSION PROBLEM

Sid has 14 coins. He has 2 fewer nickels than dimes. How many nickels and dimes does Sid have?

Solution: *8 dimes, 6 nickels*

TALK ABOUT IT

Have students talk with a partner or with a group about how they solved the Extension Problem. Students can share their different ways of thinking. Ask a question like, **How did you use a wrong guess to help you make another guess?**

WRITE YOUR OWN PROBLEM

Have students write similar problems of their own. Students can exchange problems and solve them.

PRACTICE

Similar Practice Problems: 48, 49, 50, 96

When you give students a Practice Problem, ask questions such as, **Have you solved a problem like this before? What strategies helped you solve it?**

9 Layla is at swim camp. She is learning how to to dive. Layla dives 2 times on Monday. She dives 3 times on Tuesday. She dives 4 times on Wednesday. Layla keeps diving more times each day in the same way. How many times will Layla dive on Saturday?

FIND OUT

- **What is the problem about?** Encourage students to restate the problem in their own words.

- **What do you have to find out to solve the problem?** *How many times Layla will dive on Saturday*

- **Find out what the problem tells you.**

  How many times does Layla dive on Monday? *2 times*

  How many times does she dive on Tuesday? *3 times*

  How many times does she dive on Wednesday? *4 times*

  What do you know about the way Layla keeps diving? *She keeps diving more times every day in the same way.*

CHOOSE STRATEGIES

You can *Use or Look for a Pattern* and *Use or Make a Table*. The table will help you keep track of the numbers of dives. Look for the pattern in the numbers in the table.

 TEACHING TIP

Have students look for examples of patterns around them. Help them to understand that in a pattern, something happens again and again in the same way. Sometimes a pattern can be found in a sequence of numbers that grows larger or smaller in the same way.

SOLVE IT

Look at the table that has been started.

1. **What will you keep track of in the first column?** *The days*

   What will you keep track of in the second column? *The number of times Layla dives each day*

Day	Number of Dives
Monday	2
Tuesday	3
Wednesday	4
Thursday	5
Friday	6
Saturday	7

2. **How many times does Layla dive on Monday?** *2 times* **Find that number in the table.**

3. **How many times does she dive on Tuesday?** *3 times* **Write that number in the table.**

4. **How many times does she dive on Wednesday?** *4 times* **Write that number in the table.**

5. **Look for a pattern in the numbers in the table. What pattern do you see?** Students may describe the pattern in different ways. One possible way is shown. *Each day, Layla dives 1 more time than she did the day before.*

6. **Use the pattern. Finish the table. Keep writing numbers in the table until you know how many times Layla will dive on Saturday.**

7. **How many times will Layla dive on Saturday?**

Solution: *7 times*

LOOK BACK

Students should read or listen to the problem again and check their work. Encourage them to ask themselves, **Did I answer the question that was asked in the problem? Is my answer right?**

EXTENSION PROBLEM

At the swim camp, 10 campers go diving on Monday. Nine campers go diving on Tuesday, and 8 campers go diving on Wednesday. Roland sees a pattern in the way fewer campers go diving every day. How many campers will go diving on Saturday?

Solution: *5 campers*

TALK ABOUT IT

Have students talk with a partner or small group about how they solved the Extension Problem. Students can share their different ways of thinking. Ask a question like, **How did using the table and pattern help you solve the problem?**

WRITE YOUR OWN PROBLEM

Have students write similar problems of their own. Students can then exchange problems and solve them.

PRACTICE

Similar Practice Problems: 51, 52, 53, 97

When you give students a Practice Problem, ask questions such as, **Have you solved a problem like this before? What strategies helped you solve it?**

10 Ryan watches a model train go around a track. On the first trip around, the train whistle blows 1 time. On the second trip, the whistle blows 3 times. On the third trip, the whistle blows 5 times.

The whistle keeps blowing more times on each trip in the same way. How many times will the whistle blow on the sixth trip?

FIND OUT

- **What is the problem about?** Encourage students to restate the problem in their own words.

- **What do you have to find out to solve the problem?** *How many times the train whistle will blow on the sixth trip*

- **Find out what the problem tells you.**

   How many times does the whistle blow on the first trip? *1 time*

   How many times does the whistle blow on the second trip? *3 times*

   How many times does the whistle blow on the third trip? *5 times*

   What do you know about the way the whistle keeps blowing? *It keeps blowing more times on each trip in the same way.*

CHOOSE STRATEGIES

You can *Use or Look for a Pattern* and *Use or Make a Table*. The table will help you keep track of the number of times the whistle blows. Look for a pattern in the numbers in the table.

 TEACHING TIP

Ask students to explain what a pattern is. Help them to understand that in a pattern, something happens again and again in the same way. Sometimes a pattern can be found in a sequence of numbers that grows larger or smaller in the same way.

SOLVE IT

Look at the table that has been started.

1. **What will you keep track of in the first column?** *The number of the trip*

   What will you keep track of in the second column? *The number of times the whistle blows*

Trip	Number of Times the Whistle Blows
1st	1
2nd	3
3rd	5
4th	7
5th	9
6th	11

2. **How many times does the whistle blow on the first trip?** *1 time* **Find that number in the table.**

3. **How many times does the whistle blow on the second trip?** *3 times* **Write that number in the table.**

4. **How many times does the whistle blow on the third trip?** *5 times* **Write that number in the table.**

5. **Look for a pattern in the numbers in the table. What pattern do you see?** Students may describe the pattern in different ways. One possible way is shown. *On each trip, the whistle blows 2 more times than on the trip before it.*

6. **Use the pattern. Finish the table. When can you stop?** *We can stop when we find out how many times the whistle will blow on the sixth trip.*

7. **How many times will the whistle blow on the sixth trip?**

   Solution: *11 times*

LOOK BACK

Students should read or listen to the problem again and check their work. Encourage them to ask themselves, **Did I answer the question that was asked in the problem? Is my answer right?**

EXTENSION PROBLEM

The train keeps going around the track. It keeps blowing its whistle more times in the same way. How many times will the whistle blow on the tenth trip?

Solution: *19 times*

TALK ABOUT IT

Have students talk with a partner or small group about how they solved the Extension Problem. Students can share their different ways of thinking. Ask questions like, **How did using the table and pattern help you solve the problem? Did anything else help you?**

WRITE YOUR OWN PROBLEM

Have students write similar problems of their own. Students can then exchange problems and solve them.

PRACTICE

Similar Practice Problems: 51, 52, 53, 97

When you give students a Practice Problem, ask questions such as, **Have you solved a problem like this before? What strategies helped you solve it?**

Use Logical Reasoning
Act Out or Use Objects

 Each pair needs: 4 cubes
(1 each of blue, red, yellow, and green)

11 Baby Jana lines up 4 rubber ducks in the tub. The red duck is behind the blue duck. The yellow duck is behind the red duck. The yellow duck is not next to the green duck. What color is each duck in the line?

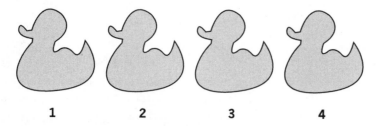

1   2   3   4

FIND OUT

• **What is the problem about?** Encourage students to restate the problem in their own words.

• **What do you have to find out to solve the problem?** *The color of each duck in the line*

• **Find out what the problem tells you.**

  How many ducks are lined up? *4 ducks*

  What colors are the ducks? *Blue, red, yellow, and green*

  What do you know about the red duck? *It is behind the blue duck.*

  What do you know about the yellow duck? *It is behind the red duck. It is not next to the green duck.*

CHOOSE STRATEGIES

You can *Use Logical Reasoning* and *Act Out or Use Objects*. Use cubes to show the colors of the ducks.

 TEACHING TIP

Introduce the language of logical thinking. This often includes conditional statements such as: "If ... then," "if something is true, then ...," or "if something is **not** true, then ...." Practice with examples such as: **If** today is Monday, **then** tomorrow is Tuesday. **If** today is Thursday, then tomorrow is **not** Sunday.

SOLVE IT

Put cubes on the ducks to show the colors.

1. **What do you know about the red duck?** *It is behind the blue duck.*

2. **If the red duck is behind the blue duck, then where can you put a red cube?** *Second in the line*

3. **What do you know about the yellow duck?** *It is behind the red duck. It is not next to the green duck.*

4. **If the yellow duck is behind the red duck, then where can you put a yellow cube?** *Third in line*

5. **If the green duck is not next to the yellow duck, then where could it be?** *Maybe it's first in line.*
Do you have to move any of your cubes? *Yes, we need to move the blue to be second, red to be third, yellow to be last.*

6. **What color is each duck in the line?**

   Solution: *1 – green, 2 – blue, 3 – red, 4 – yellow*

LOOK BACK

Students should read or listen to the problem again and check their work. Encourage them to ask themselves, **Did I answer the question that was asked? Is my answer right?**

EXTENSION PROBLEM

During her bath, Baby Jana lined up her 4 toy boats. A green boat is between a blue boat and a yellow boat. A red boat is in front of a blue boat. What color is each boat in line?

Solution: *1 – red, 2 – blue, 3 – green, 4 – yellow*

TALK ABOUT IT

Have students talk with a partner or with a group about how they solved the Extension Problem. Students can share their different ways of thinking. Ask a question like, **How did you use "if ... then" thinking to help you line up the boats?**

WRITE YOUR OWN PROBLEM

Have students write similar problems of their own. Students can exchange problems and solve them.

PRACTICE

Similar Practice Problems: 54, 55, 56

When you give students a Practice Problem, ask questions such as, **Have you solved a problem like this before? What strategies helped you solve it?**

Use Logical Reasoning
Act Out or Use Objects

 Each pair needs: 6 cubes
(2 red, 2 green, and 2 blue)

12 Chico has 4 model race cars lined up.
A blue car is between 2 red cars.
The green car is not first in line. What
color is each race car in the line?

1   2   3   4

FIND OUT

- **What is the problem about?** Encourage students
  to restate the problem in their own words.

- **What do you have to find out to solve the problem?**
  The color of each car in the line

- **Find out what the problem tells you.**

  How many race cars are lined up? *4 race cars*

  What colors are the cars? *1 blue, 2 red, 1 green*

  What do you know about the blue car? *It is
  between 2 red cars.*

  What do you know about the green car? *It is not
  first in line.*

CHOOSE STRATEGIES

You can *Use Logical Reasoning* **and** *Act Out or
Use Objects.* **Use cubes to show the cars.**

 TEACHING TIP

Introduce the language of logical thinking. This often includes
conditional statements such as: "If ... then," "if something is
true, then ...," or "if something is **not** true, then ...." Practice
with examples such as: **If** today is Monday, **then** tomorrow is
Tuesday. **If** today is Thursday, then tomorrow is **not** Sunday.

SOLVE IT

Put cubes on the cars to show the colors.

1. **What do you know about the blue car?** *It is
   between 2 red cars.*

2. **If the blue car is between the 2 red cars, then
   where can you put the blue and red cubes?**
   *The blue cube can be on the second or third car,
   in between the 2 red cubes.*

3. **What do you know about the green car?** *It is not first in line.*

4. **If the green car is not first in line, then where does it have to be?** *Behind the second red car*

5. **Where can you put a green cube?** *On the car that is last in line*

6. **What color is each car in the line?**

   Solution: *1 – red, 2 – blue, 3 – red, 4 – green*

LOOK BACK

Students should read or listen to the problem again and check their work. Encourage them to ask themselves, **Did I answer the question that was asked? Is my answer right?**

EXTENSION PROBLEM

Chico has 5 dinosaurs lined up. A blue dinosaur is between 2 yellow dinosaurs. A green dinosaur is behind a yellow dinosaur. The green dinosaur is in front of a red dinosaur. What color is each dinosaur in the line?

Solution: *1 – yellow, 2 – blue, 3 – yellow, 4 – green, 5 – red*

TALK ABOUT IT

Have students talk with a partner or with a group about how they solved the Extension Problem. Students can share their different ways of thinking. Ask a question like, **How did you use "if ... then" thinking to help you line up the dinosaurs?**

WRITE YOUR OWN PROBLEM

Have students write similar problems of their own. Students can exchange problems and solve them.

PRACTICE

Similar Practice Problems: 54, 55, 56

When you give students a Practice Problem, ask questions such as, **Have you solved a problem like this before? What strategies helped you solve it?**

Make an Organized List
Act Out or Use Objects

 Each pair needs: 3 dimes, 3 nickels, and 3 pennies

13 Tania has 3 pennies, 3 nickels, and 2 dimes. She puts her coins in groups of 2. Each group is different from every other group. What coins could be in each group?

Hint: You may find more than one answer.

FIND OUT

- **What is the problem about?** Encourage students to restate the problem in their own words.

- **What do you have to find out to solve the problem?** *What coins could be in each group*

- **Find out what the problem tells you.**

  What coins does Tania have? *3 pennies, 3 nickels, and 2 dimes*

  What does Tania do with her coins? *She puts them in groups of 2.*

  What do you know about the groups of coins? *Each group is different from every other group.*

 TEACHING TIP

To help students understand what makes a group of coins *different* from every other group, have them take 3 pennies and 3 nickels. Ask them to make 3 different groups of 2 coins. Point out that a group of 2 pennies is different from a group of 1 penny and 1 nickel. Help students understand that the order in which they put the coins does not matter. A group of 1 penny and 1 nickel is the same as a group of 1 nickel and 1 penny.

CHOOSE STRATEGIES

You can *Make an Organized List* and *Act Out or Use Objects*. Use play money to show the coins. The organized list will help you keep track of all the groups.

SOLVE IT

Use play money. Show each group of 2 coins.

1. **How many pennies will you use?** *3 pennies*

2. **How many nickels will you use?** *3 nickels*

3. **How many dimes will you use?** *2 dimes*

4. **Make different groups of 2 coins. Start with a group of 2 pennies.**

5. **Finish the list to show each group of coins.**

penny, penny
nickel, nickel
dime, dime
penny, nickel

Because there is more than one way to put the coins in different groups of 2, students may want to make more than one organized list.

Solutions:

penny – penny	*penny – penny*
nickel – nickel	*nickel – nickel*
dime – dime	*penny – dime*
penny – nickel	*nickel – dime*

LOOK BACK

Students should read or listen to the problem again and check their work. Encourage them to ask themselves, **Did I answer the question that was asked in the problem? Is my answer right?**

EXTENSION PROBLEM

Now Tania has 3 pennies, 3 nickels, 3 dimes, and 1 quarter. She puts her coins in groups of 2. Each group is different from every other group. What coins could be in each group?

Hint: You may find more than one answer.

Solutions include:

penny – penny	*penny – nickel*
nickel – nickel	*penny – dime*
dime – dime	*penny – quarter*
penny – nickel	*nickel – nickel*
dime – quarter	*dime – dime*

Students may find other solutions as well.

TALK ABOUT IT

Have students talk with a partner or small group about how they solved the Extension Problem. Students can share their different ways of thinking. Ask questions like, **How did using play money help you solve the problem? How did the organized list help you?**

WRITE YOUR OWN PROBLEM

Have students write similar problems of their own. Students can then exchange problems and solve them.

PRACTICE

Similar Practice Problems: 57, 58, 59

When you give students a Practice Problem, ask questions such as, **Have you solved a problem like this before? What strategies helped you solve it?**

Make an Organized List
Act Out or Use Objects

 Each pair needs: 10 cubes
(3 yellow, 3 green, 2 blue, and 2 red)

14 Tomas has 10 balloons for his party. He has
3 yellow balloons, 3 green balloons, 2 blue
balloons, and 2 red balloons. He ties the balloons
in groups of 2. Each group is different from
every other group. What colors are the balloons
in each group?

Hint: You may find more than one answer.

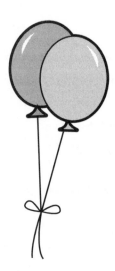

FIND OUT

- **What is the problem about?** Encourage students
to restate the problem in their own words.

- **What do you have to find out to solve the problem?**
The colors of the balloons in each group

- **Find out what the problem tells you.**

  What colors are the balloons Tomas has? *3 yellow,
  3 green, 2 blue, and 2 red*

  What does Tomas do with his balloons? *He ties
  them in groups of 2.*

  What do you know about the groups of balloons?
  *Each group of balloons is different from every
  other group.*

 TEACHING TIP

To help students understand what makes a group of 2 balloons
different from every other group, have them take 3 yellow
cubes and 3 green cubes. Ask them to make 3 different
groups of 2 cubes. Point out that a group of 2 green cubes
is different from a group of 1 green cube and 1 yellow cube.
Help students understand that the order in which they put
the cubes does not matter. A group of 1 yellow cube and
1 green cube (or balloon) is the same as a group of 1 green
cube and 1 yellow cube (or balloon).

CHOOSE STRATEGIES

**You can *Make an Organized List* and *Act Out or
Use Objects*. Use cubes to show the balloons.
The organized list will help you keep track of all
the groups.**

SOLVE IT

Use cubes to show the balloons. Show each different group of 2 balloons.

1. **How many yellow cubes will you use?** *3 yellow cubes*

2. **How many green cubes will you use?** *3 green cubes*

3. **How many blue cubes will you use?** *2 blue cubes*

4. **How many red cubes will you use?** *2 red cubes*

5. **Make different groups of 2 cubes.**

6. **Finish the list to show the colors in each group.**

yellow, yellow
green, green
blue, blue
red, red
yellow, green

Because there is more than one way to put the balloons in different groups of 2, students may want to make more than one organized list.

Solutions include:

yellow – yellow	*yellow – green*
green – green	*yellow – blue*
blue – blue	*yellow – red*
red – red	*green – green*
yellow – green	*red – blue*

Students may find other solutions as well.

LOOK BACK

Students should read or listen to the problem again and check their work. Encourage them to ask themselves, **Did I answer the question that was asked in the problem? Is my answer right?**

EXTENSION PROBLEM

Maria also has 10 balloons. She has 2 yellow balloons, 3 green balloons, 2 red balloons, and 3 blue balloons. She ties them into groups of 2 balloons. Each group is different from every other group. What colors are the balloons in each group?

Hint: You may find more than one answer.

Solutions include:

yellow – yellow	*yellow – green*
green – green	*yellow – red*
red – red	*green – red*
blue – blue	*green – blue*
green – blue	*blue – blue*

TALK ABOUT IT

Have students talk with a partner or small group about how they solved the Extension Problem. Students can share their different ways of thinking. Ask questions like, **How did using cubes help you solve the problem? How did the organized list help you?**

WRITE YOUR OWN PROBLEM

Have students write similar problems of their own. Students can then exchange problems and solve them.

PRACTICE

Similar Practice Problems: 57, 58, 59

When you give students a Practice Problem, ask questions such as, **Have you solved a problem like this before? What strategies helped you solve it?**

Make It Simpler
Make an Organized List
Act Out or Use Objects

 Each pair needs: 5 cubes (any colors)

15 Reggie has 2 pockets in his pants. He puts 5 stones into his pockets. He puts at least 1 stone into each pocket. What are all the different ways that Reggie could put the stones into his pockets?

 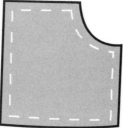

A   B

FIND OUT

- **What is the problem about?** Encourage students to restate the problem in their own words.

- **What do you have to find out to solve the problem?** *All the different ways Reggie could put the stones into his pockets*

- **Find out what the problem tells you.**

  How many pockets does Reggie have? *2 pockets*

  What does he put into his pockets? *5 stones*

  What do you know about the way he puts the stones into his pockets? *He puts at least 1 into each pocket.*

CHOOSE STRATEGIES

You can *Make It Simpler, Make An Organized List,* and *Act Out Or Use Objects* to help you solve this problem. Use cubes to show the stones.

🍎 TEACHING TIP

Talk with students about what Make It Simpler means. In this problem, for example, Reggie is putting 5 stones into his 2 pockets. To make it simpler, students can try putting 3 stones into the 2 pockets. Doing a simpler problem helps them understand how to do the larger problem.

SOLVE IT

Put cubes into the pockets to show the stones.

1. **Make it simpler. Think of 3 stones. Use 3 cubes. Put the cubes into pocket A and pocket B.**

2. **How many cubes are in pocket A?** *1 cube*
   Then how many cubes are in pocket B? *2 cubes*
   Write that number in the list.

3. **Now put the cubes into the pockets in a different way. How many cubes are in pocket A?** *2 cubes* **How many cubes are in pocket B?** *1 cube* **Write those numbers in the list.**

4. **Is there another way to put the 3 cubes into the pockets?** *No.*

A	B
1	2
2	1

5. **Now use 5 cubes. Finish the list. Show all the different ways that Reggie could put 5 stones into his pockets.**

A	B
1	4
2	3
3	2
4	1

Solution: *See list above.*

LOOK BACK

Students should read or listen to the problem again and check their work. Encourage them to ask themselves, **Did I answer the question that was asked? Is my answer right?**

EXTENSION PROBLEM

Reggie found another stone, so now he has 6 stones. If he puts at least 1 stone into each pocket, what are all the different ways that Reggie could put 6 stones into his pockets?

Solution:

A	B
1	5
2	4
3	3
4	2
5	1

TALK ABOUT IT

Have students talk with a partner or with a group about how they solved the Extension Problem. Students can share their different ways of thinking. Ask questions like, **How did making it simpler help you solve this problem? How did the organized list help you?**

WRITE YOUR OWN PROBLEM

Have students write similar problems of their own. Students can exchange problems and solve them.

PRACTICE

Similar Practice Problems: 60, 61, 62

When you give students a Practice Problem, ask questions such as, **Have you solved a problem like this before? What strategies helped you solve it?**

**Each pair needs: 6 cubes
(any colors)**

16 Elena has 6 fish. She is putting the fish
into 2 bowls. She puts at least 1 fish into
each bowl. What are all the different ways
that Elena could put the fish into the bowls?

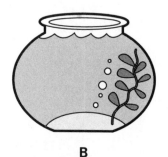

A   B

FIND OUT

• **What is the problem about?** Encourage students
to restate the problem in their own words.

• **What do you have to find out to solve the problem?**
*All the different ways Elena could put the fish into
the bowls*

• **Find out what the problem tells you.**

How many fish does Elena have? *6 fish*

How many bowls does Elena have? *2 bowls*

**What do you know about how she is putting the
fish into the bowls?** *She puts at least 1 fish into
each bowl.*

CHOOSE STRATEGIES

You can *Make It Simpler, Make an Organized
List,* and *Act Out or Use Objects.* Use cubes to
show the fish.

 TEACHING TIP

Talk with students about what Make It Simpler means.
In this problem, for example, Elena is putting 6 fish into
2 bowls. To make it simpler, students can try putting 4 fish
into the 2 bowls. Doing a simpler problem helps them
understand how to do the larger problem.

SOLVE IT

Put cubes into the bowls to show the fish.

1. **Make it simpler. Think of 4 fish. Use 4 cubes.
Put the cubes into bowl A and bowl B.** Students
may choose different ways to place the cubes.
The following example is based on first putting
1 fish into bowl A and 3 fish into bowl B, and
then putting 2 fish into bowl A and 2 into bowl B.

2. **How many cubes are in bowl A?** *1 cube* **How many
cubes are in bowl B?** *3 cubes* **Write that number
in the list.**

3. **Put the cubes into the bowls in a different way. How many cubes are in bowl A?** *2 cubes* **How many cubes are in bowl B?** *2 cubes* **Write these numbers in the list.**

4. **Is there another way to put the cubes into the bowls?** *Yes, 3 in bowl A and 1 in bowl B* **Write those numbers in the list.**

A	B
1	3
2	2
3	1

5. **Now use 6 cubes. Finish the list. Show all the different ways that Elena could put 6 fish into 2 bowls.**

A	B
1	5
2	4
3	3
4	2
5	1

Solution: *See list above.*

LOOK BACK

Students should read or listen to the problem again and check their work. Encourage them to ask themselves, **Did I answer the question that was asked? Is my answer right?**

EXTENSION PROBLEM

Elena gets 1 more fish, so now she has 7 fish. If she puts at least 1 fish into each bowl, what are all the different ways that Elena could put her fish into 2 bowls?

Solution:

A	B
1	6
2	5
3	4
4	3
5	2
6	1

TALK ABOUT IT

Have students talk with a partner or with a group about how they solved the Extension Problem. Students can share their different ways of thinking. Ask questions like, **How did making it simpler help you solve this problem? How did the organized list help you?**

WRITE YOUR OWN PROBLEM

Have students write similar problems of their own. Students can exchange problems and solve them.

PRACTICE

Similar Practice Problems: 60, 61, 62

When you give students a Practice Problem, ask questions such as, **Have you solved a problem like this before? What strategies helped you solve it?**

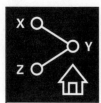

Use or Make a Picture or Diagram
Act Out or Use Objects

 Each pair needs: 2 cubes of different colors

17 Sue and Willy are playing a board game. Sue's game marker is on START. She moves her marker forward 5 spaces on the track. Next, she moves her marker back 3 spaces. Next, she moves her marker forward 8 spaces. Willy's marker is on space 9. Who is ahead, Sue or Willy?

FIND OUT

- **What is the problem about?** Encourage students to restate the problem in their own words.

- **What do you have to find out to solve the problem?** *Who is ahead on the game board, Sue or Willy?*

- **Find out what the problem tells you.**

  Where is Sue's marker when she begins the game? *Her marker is on START.*

  How many spaces does she move her marker forward? *5 spaces*

  How many spaces does she move her marker back? *3 spaces*

  Then how many spaces does she move her marker forward? *8 spaces*

  What space is Willy's marker on? *Space 9*

CHOOSE STRATEGIES

You can *Use or Make a Picture or Diagram* and *Act Out or Use Objects*. The picture or diagram will help you keep track of the spaces the players' markers are on.

SOLVE IT

Use cubes for the game markers. Move the cubes on the track.

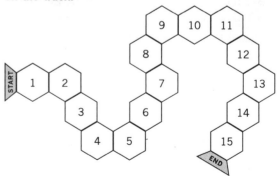

1. **Where is Sue's marker when she begins the game?** *On* START
2. **How many spaces does Sue move her marker forward?** *5 spaces*
3. **Then what space is Sue's marker on?** *Space 5*
4. **How many spaces does Sue move her marker back?** *3 spaces*
5. **Then what space is Sue's marker on?** *Space 2*
6. **How many spaces does Sue move her marker forward?** *8 spaces*
7. **Then what space is Sue's marker on?** *Space 10*
8. **What space is Willy's marker on?** *Space 9*
9. **Who is ahead, Sue or Willy?**

   Solution: *Sue is ahead by 1 space.*

LOOK BACK

Students should read or listen to the problem again and check their work. Encourage them to ask themselves, **Did I answer the question that was asked in the problem? Is my answer right?**

EXTENSION PROBLEM

Now Dan and Rosa are playing the game. Dan's marker is on START**. He moves his marker forward 7 spaces. Next, he moves his marker back 4 spaces. Next, he moves his marker forward 9 spaces. Rosa's marker is on space 12. Who is ahead, Dan or Rosa?**

Solution: *Neither player, because both of their markers are on space 12.*

TALK ABOUT IT

Have students talk with a partner or small group about how they solved the Extension Problem. Students can share their different ways of thinking. Ask a question like, **How did you use the diagram to help you solve the problem?**

WRITE YOUR OWN PROBLEM

Have students write similar problems of their own. Students can then exchange problems and solve them.

PRACTICE

Similar Practice Problems: 63, 64, 65

When you give students a Practice Problem, ask questions such as, **Have you solved a problem like this before? What strategies helped you solve it?**

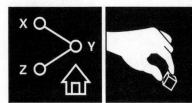

**Each pair needs: 1 cube
(any color)**

18 Max and his friends are sitting on the school steps. Max is on the first step. He goes up 2 steps and talks with Tim. He goes up 3 more steps and talks with Ana. Next he goes down 2 steps and talks with Ben. Then he goes up 5 steps and talks with Dom. What step is Dom on?

FIND OUT

- **What is the problem about?** Encourage students to restate the problem in their own words.

- **What do you have to find out to solve the problem?** *What step Dom is on*

- **Find out what the problem tells you.**

  Where does Max start out? *On the first step*

  How many steps does Max go up to talk with Tim? *2 steps*

  How many steps does Max go up to talk with Ana? *3 steps*

  How many steps does Max go down to talk with Ben? *2 steps*

  How many steps does Max go up to talk with Dom? *5 steps*

CHOOSE STRATEGIES

You can *Use or Make a Picture or Diagram* and *Act Out or Use Objects*. Using a cube on the diagram will help you keep track of the steps where Max talks with his friends.

SOLVE IT

Use a cube to show where Max goes. Add steps when you need to.

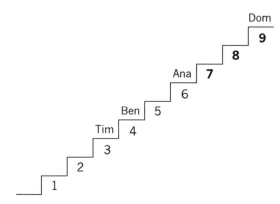

1. **Where does Max start out?** *On the first step*
2. **How many steps does Max go up to talk with Tim?** *2 steps*
3. **Then what step is Tim on?** *Step 3* **Write Tim's name on that step.**
4. **How many steps does Max go up to talk with Ana?** *3 steps*
5. **Then what step is Ana on?** *Step 6* **Write Ana's name on that step.**
6. **How many steps does Max go down to talk with Ben?** *2 steps*
7. **Then what step is Ben on?** *Step 4* **Write Ben's name on that step.**
8. **How many steps does Max go up to talk with Dom?** *5 steps*
9. **What step is Dom on?**

   Solution: *Dom is on step 9.*

LOOK BACK

Students should read or listen to the problem again and check their work. Encourage them to ask themselves, **Did I answer the question that was asked in the problem? Is my answer right?**

EXTENSION PROBLEM

Leslie starts on the first step. She goes up 4 steps and talks with Raj. She goes down 3 steps and talks with Ari. Then she goes up 8 steps and talks with Dan. What steps are Raj, Ari, and Dan on?

Solution: *Raj – step 5, Ari – step 2, Dan – step 10*

TALK ABOUT IT

Have students talk with a partner or small group about how they solved the Extension Problem. Students can share their different ways of thinking. Ask a question like, **How did you use the diagram to help you solve the problem?**

WRITE YOUR OWN PROBLEM

Have students write similar problems of their own. Students can then exchange problems and solve them.

PRACTICE

Similar Practice Problems: 63, 64, 65

When you give students a Practice Problem, ask questions such as, **Have you solved a problem like this before? What strategies helped you solve it?**

Each pair needs: 2 of each coin (half dollars, quarters, dimes, nickels, and pennies)

19 Ariel bought some worms at Franco's Fishing Store. She paid more than 8 cents and less than 17 cents. She paid with 2 pennies and 2 other coins. How much did Ariel pay for the worms?

FIND OUT

- **What is the problem about?** Encourage students to restate the problem in their own words.

- **What do you have to find out to solve the problem?** *How much Ariel paid for the worms*

- **Find out what the problem tells you.**

  What do you know about how much Ariel paid? *She paid more than 8 cents and less than 17 cents.*

  What coins did she pay with? *2 pennies and 2 other coins*

🍎 **TEACHING TIP**

You may want to review the values for a half dollar, quarter, dime, nickel, and penny. You may also want to review what *more than* 8 cents and *less than* 17 cents means.

CHOOSE STRATEGIES

You can *Use or Make a Table* and *Act Out or Use Objects*. Use play money to show the coins.

SOLVE IT

Use play money. Show different amounts.

1. **What numbers in the table would not be possible?** *8 and 17* **Cross out the numbers that could not be the answer.**

cents	8	9	10	11	12	13	14	15	16	17

2. **What numbers in the table could be the answer?** *9, 10, 11, 12, 13, 14, 15, 16*

3. **What coins do you know Ariel used?** *2 pennies and 2 other coins*

   What other kinds of coins could Ariel have used? *Half dollars, quarters, dimes, or nickels*

   Could one of the 2 coins be a half dollar? *No, that would make the total over 50 cents.*

   Answers to the following questions will vary, depending on which 2 other coins the student chooses. The sample answers shown here are based on a choice of 1 dime and 1 nickel.

4. **Choose 2 other coins that Ariel could have used. What coins did you choose?** *1 dime and 1 nickel*

5. **Add 2 pennies to your 2 coins. What is the total value of your coins and the 2 pennies?** *17 cents*

6. **Is your answer right?** *No.*

7. **If your answer is not right, choose another 2 coins. Keep doing this until your coins together equal a number in the table.**

8. **How much did Ariel pay for the worms?**

   Solution: *12 cents (2 pennies, 2 nickels)*

cents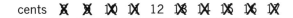

LOOK BACK

Students should read or listen to the problem again and check their work. Encourage them to ask themselves, **Did I answer the question that was asked? Is my answer right?**

EXTENSION PROBLEM

Brenda bought some fishing worms, too. She paid more than 14 cents and less than 24 cents. She paid with 4 pennies and 2 other coins. How much did Brenda pay for her worms?

Solution: *19 cents (4 pennies, 1 nickel, 1 dime)*

TALK ABOUT IT

Have students talk with a partner or with a group about how they solved the Extension Problem. Students can share their different ways of thinking. Ask a question like, **How did you decide which 2 coins to choose?**

WRITE YOUR OWN PROBLEM

Have students write similar problems of their own. Students can exchange problems and solve them.

PRACTICE

Similar Practice Problems: 66, 67, 68

When you give students a Practice Problem, ask questions such as, **Have you solved a problem like this before? What strategies helped you solve it?**

Use or Make a Table
Act Out or Use Objects

 Each pair needs: 2 of each coin (half dollars, quarters, dimes, nickels, and pennies)

20 Derrick bought a ticket for the Haunted House. He paid less than 26 cents and more than 16 cents. He paid with 2 nickels and 2 other coins. How much did Derrick pay?

FIND OUT

- **What is the problem about?** Encourage students to restate the problem in their own words.

- **What do you have to find out to solve the problem?** *How much Derrick paid for his Haunted House ticket*

- **Find out what the problem tells you.**

  What do you know about how much Derrick paid? *He paid less than 26 cents and more than 16 cents.*

  What coins did he pay with? *2 nickels and 2 other coins*

 TEACHING TIP

You may want to review the values for a half dollar, quarter, dime, nickel, and penny. You may also want to review what *less than* 26 cents and *more than* 16 cents means.

CHOOSE STRATEGIES

You can *Use or Make a Table* and *Act Out or Use Objects*. Use play money to show the coins.

SOLVE IT

Look at the table that has been started.

cents	16	17	**18**	**19**	**20**	**21**	**22**	**23**	**24**	**25**	26

1. **Write the missing numbers in the table.**

2. **What numbers in the table would not be possible?** *16 and 26* **Cross out the numbers that could not be the answer.**

3. **What numbers in the table could be the answer?** *17, 18, 19, 20, 21, 22, 23, 24, 25*

   Answers to the following questions will vary, depending on which 2 other coins the student chooses. The sample answers shown here are based on a choice of 2 dimes.

4. **What coins do you know Derrick used?** *2 nickels and 2 other coins* **What other kinds of coins could Derrick have used?** *Half dollars, quarters, dimes, or pennies*

   Could one of the 2 coins be a half dollar? *No, that would make the total over 50 cents.*

40   Problem Solver II

5. **Use play money to show different possible amounts. Choose 2 other coins Derrick could have used. What coins did you choose?** *2 dimes*

6. **Add 2 nickels to the 2 coins. What is the total value of your coins and the 2 nickels?** *30 cents*

7. **Is your answer right?** *No.*

8. **If your answer is not right, choose another 2 coins. Keep doing this until your coins together equal a number in the table.**

9. **How much did Derrick pay?**

   Solution: *21 cents (2 nickels, 1 penny, 1 dime)*

   cents 1̶6̶ 1̶7̶ 1̶8̶ 1̶9̶ 2̶0̶ 21 2̶2̶ 2̶3̶ 2̶4̶ 2̶5̶ 2̶6̶

LOOK BACK

Students should read or listen to the problem again and check their work. Encourage them to ask themselves, **Did I answer the question that was asked? Is my answer right?**

EXTENSION PROBLEM

Derrick is at the Fall Fair. He bought some pumpkin seeds. He paid less than 31 cents and more than 21 cents. He paid with 1 dime and 3 other coins. How much did Derrick pay?

Solution: *25 cents (1 dime, 3 nickels)*

TALK ABOUT IT

Have students talk with a partner or with a group about how they solved the Extension Problem. Students can share their different ways of thinking. Ask a question like, **What helped you decide which 3 coins to choose?**

WRITE YOUR OWN PROBLEM

Have students write similar problems of their own. Students can exchange problems and solve them.

PRACTICE

Similar Practice Problems: 66, 67, 68

When you give students a Practice Problem, ask questions such as, **Have you solved a problem like this before? What strategies helped you solve it?**

21 Manuel Mouse smells cheese and peanut butter! He starts at the mouse hole. He takes this path to find a snack:
- **3 steps to the right**
- **3 steps up**
- **2 steps to the right**
- **2 steps up**

Which snack does Manuel find?

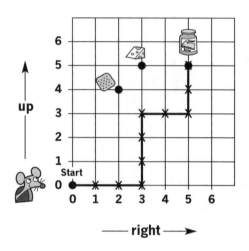

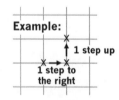

Example:

1 step up

1 step to the right

FIND OUT

- **What is the problem about?** Encourage students to restate the problem in their own words.

- **What do you have to find out to solve the problem?** *Which snack Manuel finds*

- **Find out what the problem tells you.**

  What snacks does Manuel smell? *Cheese and peanut butter*

  What do you know about the path Manuel takes? *Manuel starts at the mouse hole. He takes 3 steps to the right. Then he takes 3 steps up. Then he takes 2 steps to the right. Then he takes 2 steps up.*

TEACHING TIP

Talk about the example with the students. Show them the arrows that direct them to the right and up. Point out the length of one step in each direction. Then have them trace this path on the map with a finger: Begin at the mouse hole, which is at *Start*. Go 2 steps to the right, then 4 steps up. Ask them what is at the end of their path. *(a cracker)*

CHOOSE STRATEGIES

You can *Use or Make a Picture or Diagram*. A map is a kind of diagram. Use the map to help you follow Manuel's path.

SOLVE IT

Look at the map.

1. **Where does Manuel start?** *At the mouse hole*

2. **Draw an X on the map to show each of Manuel's steps.**

   How many steps does Manuel take to the right?
   3 steps

   How many steps does Manuel take up? *3 steps*

   Then how many steps does he take to the right?
   2 steps

   Then how many steps does he take up? *2 steps*

3. **Which snack does Manuel find?**

   Solution: *Peanut butter*

LOOK BACK

Students should read or listen to the problem again and check their work. Encourage them to ask themselves, **Did I answer the question that was asked in the problem? Is my answer right?**

EXTENSION PROBLEM

Now Millie Mouse comes out of the same mouse hole. She takes a different path to the peanut butter. She takes the same total number of steps as Manuel does. What path could Millie take?

Hint: You may find more than one answer.

Solutions include: *1 step to the right, 1 step up, 4 steps to the right, 4 steps up; or 2 steps to the right, 2 steps up, 3 steps to the right, 3 steps up*

Any path that has a total of 5 steps to the right and a total of 5 steps up will be correct, as long as it is different from the path Manuel took.

TALK ABOUT IT

Have students talk with a partner or small group about how they solved the Extension Problem. Students can share their different ways of thinking. Ask a question like, **How did you use the map to help you solve the problem?**

WRITE YOUR OWN PROBLEM

Have students write similar problems of their own. Students can then exchange problems and solve them.

PRACTICE

Similar Practice Problems: 69, 70, 71

When you give students a Practice Problem, ask questions such as, **Have you solved a problem like this before? What strategies helped you solve it?**

Use or Make a Picture or Diagram

22 Polly Possum likes to eat grapes and dog food! She starts at the log. She takes this path to find her dinner:

- **2 steps to the right**
- **5 steps up**
- **4 steps to the right**
- **1 step up**

What does Polly find for dinner?

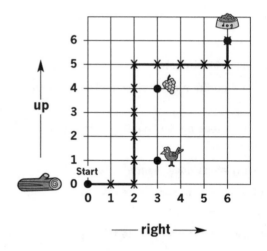

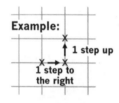

Example:

1 step up

1 step to the right

FIND OUT

- **What is the problem about?** Encourage students to restate the problem in their own words.

- **What do you have to find out to solve the problem?** *What Polly finds for dinner*

- **Find out what the problem tells you.**

  What does Polly like to eat? *Grapes and dog food*

  What do you know about the path that Polly takes? *She starts at the log. She takes 2 steps to the right. Then she takes 5 steps up. Then she takes 4 steps to the right. Then she takes 1 step up.*

 TEACHING TIP

Talk about the example with the students. Show them the arrows that direct them to the right and up. Point out the length of one step in each direction. Then have them trace this path on the map with a finger: Begin at the log, which is at *Start*. Go 3 steps to the right, then 1 step up. Ask them what is at the end of their path. *(a bird)*

CHOOSE STRATEGIES

You can *Use or Make a Picture or Diagram*. A map is a kind of diagram. Use the map to help you follow Polly's path.

SOLVE IT

Look at the map.

1. **Where does Polly start?** *At the log*

2. **Draw an X on the map to show each of Polly's steps.**

   How many steps does Polly take to the right?
   2 steps

   How many steps does Polly take up? *5 steps*

   Then how many steps does she take to the right?
   4 steps

   Then how many steps does she take up? *1 step*

3. **What does Polly find for dinner?**

   Solution: *Dog food*

LOOK BACK

Students should read or listen to the problem again and check their work. Encourage them to ask themselves, **Did I answer the question that was asked in the problem? Is my answer right?**

EXTENSION PROBLEM

Patrick Possum also starts at the log. He takes a different path than Polly took, but he takes the same total number of steps. He wants both grapes and dog food for dinner. What path could Patrick take?

Hint: You may find more than one answer.

Solutions include: *2 steps to the right, 1 step up, 1 step to the right, 3 steps up, 1 step to the right, 2 steps up, 2 steps to the right; or 3 steps to the right, 4 steps up, 3 steps to the right, 2 steps up*

TALK ABOUT IT

Have students talk with a partner or small group about how they solved the Extension Problem. Students can share their different ways of thinking. Ask a question like, **How did you use the map to help you solve the problem?**

WRITE YOUR OWN PROBLEM

Have students write similar problems of their own. Students can then exchange problems and solve them.

PRACTICE

Similar Practice Problems: 69, 70, 71

When you give students a Practice Problem, ask questions such as, **Have you solved a problem like this before? What strategies helped you solve it?**

23 Miss O'Day's class voted for their favorite kinds of fruit.
The graph shows how many children voted for each fruit.
• One more girl than boy voted for Angela's favorite fruit.
• The same number of girls as boys voted for Shane's favorite fruit.
• Two more boys than girls voted for Lorie's favorite fruit.

Which fruits did Angela, Shane, and Lorie vote for?

Our Favorite Fruits

Fruit	Number of Girls and Boys
apple	😊 😊 😊 😊 😊 😊 😊 😊
orange	😊 😊 😊 😊 😊 😊
banana	😊 😊 😊 😊 😊 😊 😊 😊 😊 😊 😊

😊 **is 1 boy** 😊 **is 1 girl**

FIND OUT

• **What is the problem about?** Encourage students to restate the problem in their own words.

• **What do you have to find out to solve the problem?** *Which fruits Angela, Shane, and Lorie voted for*

• **Find out what the problem tells you.**

What do you know about Angela's favorite fruit?
One more girl than boy voted for it.

What do you know about Shane's favorite fruit?
The same number of girls as boys voted for it.

What do you know about Lorie's favorite fruit?
Two more boys than girls voted for it.

CHOOSE STRATEGIES

You can *Use Logical Reasoning* and *Use or Make a Picture or Diagram*. The picture graph is a kind of diagram. Use both strategies to help you find out what fruit each child voted for.

SOLVE IT

1. **Look at the picture graph.** Encourage students to talk about the picture graph and what it shows. Review what each picture stands for.

   What are the names on the side? *3 fruits: apple, orange, banana*

   What are the pictures next to the names? *The faces of girls and boys*

   How many votes does each face show? *1 vote*

2. **Begin with the clue about Angela's favorite fruit.**
(One more girl than boy voted for this fruit.)

How many girls voted for an apple? *3 girls*
How many boys? *5 boys*

How many girls voted for an orange? *3 girls*
How many boys? *3 boys*

How many girls voted for a banana? *6 girls*
How many boys? *5 boys*

Which fruit fits the clue for Angela? *Banana*

3. **Now use the clue about Shane's favorite fruit.**
(The same number of girls as boys voted for Shane's fruit.)

Which fruit fits the clue for Shane? *Orange*

4. **Now use the clue about Lorie's favorite fruit.**
(Two more boys than girls voted for Lorie's fruit.)

Which fruit fits the clue for Lorie? *Apple*

5. **Which fruits did Angela, Shane, and Lorie vote for?**

Solution: *Angela – banana, Shane – orange,
Lorie – apple*

LOOK BACK

Students should read or listen to the problem
again and check their work. Encourage them to
ask themselves, **Did I answer the question that
was asked? Is my answer right?**

EXTENSION PROBLEM

**Megan voted for a fruit that got 3 more votes than
the apple. Bob voted for a fruit that got 5 fewer
votes than the banana. Coco voted for a fruit that
got 2 fewer votes than the apple. Which fruits did
Megan, Bob, and Coco vote for?**

Solution: *Megan – banana, Bob – orange,
Coco – orange*

TALK ABOUT IT

Have students talk with a partner or with a
group about how they solved the Extension
Problem. Students can share their different ways
of thinking. Ask a question like, **How do picture
graphs help you to get information?**

WRITE YOUR OWN PROBLEM

Have students write similar problems of their
own. Students can exchange problems and
solve them.

PRACTICE

Similar Practice Problems: 72, 73, 74, 99

When you give students a Practice Problem, ask
questions such as, **Have you solved a problem like
this before? What strategies helped you solve it?**

Use Logical Reasoning
Use or Make a Picture or Diagram

24 Mrs. James asked her class which card game they like best. The graph shows how many students voted for each game.
- Joey likes the game that got more votes than Crazy 8s.
- Maria likes the game that got fewer votes than Crazy 8s.
- Beth likes the game that got an even number of votes.

Which card games do Joey, Maria, and Beth like best?

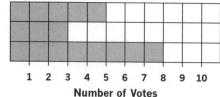

Our Favorite Card Games

Crazy 8s
Go Fish
Match It

1 2 3 4 5 6 7 8 9 10
Number of Votes

FIND OUT

- **What is the problem about?** Encourage students to restate the problem in their own words.

- **What do you have to find out to solve the problem?** *Which card games Joey, Maria, and Beth like best*

- **Find out what the problem tells you.**

  What do you know about Joey's favorite game? *It got more votes than Crazy 8s.*

  What do you know about Maria's favorite game? *It got fewer votes than Crazy 8s.*

  What do you know about Beth's favorite game? *It got an even number of votes.*

CHOOSE STRATEGIES

You can *Use Logical Reasoning* and *Use or Make a Picture or Diagram.* A bar graph is a kind of diagram. Use both strategies to help you find out which card games Joey, Maria, and Beth like best.

SOLVE IT

1. **Look at the bar graph.** Encourage students to talk about the bar graph and what it shows. Review the different parts of a bar graph.

   What are the names on the side? *3 different card games: Crazy 8s, Go Fish, Match It*

   What do the numbers at the bottom show? *The number of votes the games got*

2. **Begin with the clue about Joey's favorite game.** (It got more votes than Crazy 8s.)

   How many games got more votes than Crazy 8s?
   1 game

   If there is only 1, then what game did Joey like?
   Match It

3. **Now look at the clue for Maria's favorite game.** (It got fewer votes than Crazy 8s.)

   How many games got fewer votes than Crazy 8s?
   1 game

   If there is only 1, then what game did Maria like?
   Go Fish

4. **Now look at the clue for Beth's favorite game.** (It got an even number of votes.)

   How many games got an even number of votes?
   1 game

   If there is only 1, then what game did Beth like?
   Match It

5. **Which card games do Joey, Maria, and Beth like best?**

   Solution: *Joey – Match It, Maria – Go Fish, Beth – Match It*

LOOK BACK

Students should read or listen to the problem again and check their work. Encourage them to ask themselves, **Did I answer the question that was asked? Is my answer right?**

EXTENSION PROBLEM

Nikki voted for the game that got 3 more votes than Crazy 8s. John voted for the game that got 2 fewer votes than Crazy 8s. Troy voted for the game that got 3 fewer votes than Match It. Which card games did Nikki, John, and Troy vote for?

Solution: *Nikki – Match It, John – Go Fish, Troy – Crazy 8s*

TALK ABOUT IT

Have students talk with a partner or with a group about how they solved the Extension Problem. Students can share their different ways of thinking. Ask questions like, **How did you use the graph to solve the problem? How did you use "if ... then" thinking?**

WRITE YOUR OWN PROBLEM

Have students write similar problems of their own. Students can exchange problems and solve them.

PRACTICE

Similar Practice Problems: 72, 73, 74, 99

When you give students a Practice Problem, ask questions such as, **Have you solved a problem like this before? What strategies helped you solve it?**

**Each pair needs: 30 cubes
(10 red, 10 blue, and 10 yellow)**

25 Sheena grew some flowers. She has
3 more red flowers than blue flowers.
She has 5 more blue flowers than yellow
flowers. She has 2 yellow flowers. How
many red flowers does Sheena have?

FIND OUT

- **What is the problem about?** Encourage students
to restate the problem in their own words.

- **What do you have to find out to solve the problem?**
How many red flowers Sheena has

- **Find out what the problem tells you.**

**What do you know about the number of red
flowers that Sheena has?** *She has 3 more red
flowers than blue flowers.*

**What do you know about the number of blue
flowers that Sheena has?** *She has 5 more blue
flowers than yellow flowers.*

How many yellow flowers does Sheena have?
2 yellow flowers

CHOOSE STRATEGIES

**You can *Work Backwards* and *Act Out or Use
Objects*. Use cubes to show the flowers.**

 TEACHING TIP

Talk with students about what Working Backwards means.
Ask what they think it means. Do they ever have to work
backwards in their daily life? For example, they might have
to be in bed at a certain time. Then what time do they have
to start getting ready for bed?

SOLVE IT

1. **How many yellow flowers does Sheena have?**
2 yellow flowers **Write this number in the box
for yellow.**

   **Where in the problem is the information about the
yellow flowers?** *At the end*

2. **Work backwards from the end of the problem.
What do you know about the number of blue
flowers that Sheena has?** *She has 5 more blue
flowers than yellow flowers.*

red	blue	yellow
10	**7**	**2**

3. **If you know how many yellow flowers Sheena has, then how can you find out how many blue flowers Sheena has?** *Add 5 to 2.*

4. **How many blue flowers does Sheena have?** *7 blue flowers* **Write this number in the box for blue.**

5. **Work backwards again.**

   What do you know about the number of red flowers that Sheena has? *She has 3 more red flowers than blue flowers.*

6. **If you know how many blue flowers Sheena has, then how can you find out how many red flowers she has?** *Add 3 to 7.*

7. **How many red flowers does Sheena have?**

   Solution: *10 red flowers*

LOOK BACK

Students should read or listen to the problem again and check their work. Encourage them to ask themselves, **Did I answer the question that was asked? Is my answer right?**

EXTENSION PROBLEM

Molly also grew some flowers. She has 2 more orange flowers than pink flowers. She has the same number of pink flowers as yellow flowers. She has 5 yellow flowers. How many orange flowers does Molly have?

Solution: *7 orange flowers*

TALK ABOUT IT

Have students talk with a partner or with a group about how they solved the Extension Problem. Students can share their different ways of thinking. Ask a question like, **What helped you the most when you were working backwards?**

WRITE YOUR OWN PROBLEM

Have students write similar problems of their own. Students can exchange problems and solve them.

PRACTICE

Similar Practice Problems: 75, 76, 77

When you give students a Practice Problem, ask questions such as, **Have you solved a problem like this before? What strategies helped you solve it?**

**Each pair needs: 30 cubes
(10 blue, 10 green, and 10 red)**

26 Carlos is sorting puzzle pieces. He has
4 more blue pieces than green pieces.
He has 5 more green pieces than red
pieces. He has 1 red piece. How many
blue puzzle pieces does Carlos have?

FIND OUT

- **What is the problem about?** Encourage students
to restate the problem in their own words.

- **What do you have to find out to solve the problem?**
How many blue puzzle pieces Carlos has

- **Find out what the problem tells you.**

  **What do you know about the number of blue
  pieces that Carlos has?** *He has 4 more blue pieces
  than green pieces.*

  **What do you know about the number of green
  pieces that Carlos has?** *He has 5 more green
  pieces than red pieces.*

  How many red pieces does Carlos have? *1 red
  piece*

CHOOSE STRATEGIES

You can *Work Backwards* and *Act Out or Use
Objects.* Use cubes to show the puzzle pieces.

 TEACHING TIP

Talk with students about what Working Backwards means.
Ask what they think it means. Do they ever have to work
backwards in their daily life? For example, they might have
to be in bed at a certain time. Then what time do they have
to start getting ready for bed?

SOLVE IT

1. **How many red pieces does Carlos have?** *1 red
piece* **Write this number in the box for red.
Where in the problem is the number of red pieces
given?** *At the end*

2. **Work backwards from the end of the problem.
What do you know about the number of green
pieces that Carlos has?** *He has 5 more green
pieces than red pieces.*

blue	green	red
10	6	1

3. **If you know how many red pieces Carlos has, then how can you find out how many green pieces he has?** *Add 5 to 1.*

4. **How many green pieces does he have?** *6 green pieces* **Write this number in the box for green.**

5. **Work backwards again. What do you know about the number of blue pieces that Carlos has?** *He has 4 more blue pieces than green pieces.*

6. **If you know how many green pieces Carlos has, then how can you find out how many blue pieces he has?** *Add 4 to 6.*

7. **How many blue puzzle pieces does Carlos have?**

   Solution: *10 blue puzzle pieces*

LOOK BACK

Students should read or listen to the problem again and check their work. Encourage them to ask themselves, **Did I answer the question that was asked? Is my answer right?**

EXTENSION PROBLEM

Sonya is sorting the pieces of a different puzzle. She has 3 fewer yellow pieces than brown pieces. She has 4 fewer brown pieces than black pieces. She has 12 black pieces. How many yellow puzzle pieces does Sonya have?

Solution: *5 yellow pieces*

TALK ABOUT IT

Have students talk with a partner or with a group about how they solved the Extension Problem. Students can share their different ways of thinking. Ask a question like, **What helped you the most when you were working backwards?**

WRITE YOUR OWN PROBLEM

Have students write similar problems of their own. Students can exchange problems and solve them.

PRACTICE

Similar Practice Problems: 75, 76, 77

When you give students a Practice Problem, ask questions such as, **Have you solved a problem like this before? What strategies helped you solve it?**

27 On Saturday Anita plays a game with her sister. They play for 1 hour. Next Anita reads books for 1 hour. Then she walks her dog Bingo for 1 hour. She and Bingo get home at 4 o'clock. What time was it when Anita started playing a game with her sister?

FIND OUT

- **What is the problem about?** Encourage students to restate the problem in their own words.

- **What do you have to find out to solve the problem?** *What time it was when Anita started playing a game with her sister*

- **Find out what the problem tells you.**

  How long did Anita play the game? *1 hour*

  How long did Anita read books? *1 hour*

  How long did Anita walk her dog? *1 hour*

  What time was it when she got home? *4 o'clock*

CHOOSE STRATEGIES

You can *Work Backwards* and *Use or Make a Picture or Diagram*. Use a picture of a clock face. Use both strategies to help you find out when Anita started playing a game with her sister.

🍎 **TEACHING TIP**

Tell students they will be working backwards in two ways in this problem. They will begin at the end of the problem and work backwards through the problem. They will also work backwards on the clock face. Have students practice going backwards on the clock. Ask them what time would be 1 hour before 10 o'clock.

SOLVE IT

Use the clock to find the times.

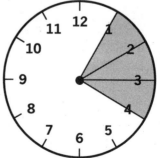

Write the times in the boxes.

Start game	Start books	Start walk	Get home
1	**2**	**3**	**4**
o'clock	o'clock	o'clock	o'clock

1. **What time was it when Anita and her dog got home?** *4 o'clock* **Where is this time given in the problem?** *At the end*

2. **Work backwards from the end of the problem. How long did Anita and her dog walk?** *1 hour*

3. **Work backwards on the clock. What time was it when they started walking?** *3 o'clock* **Write this time in the box where it belongs.**

4. **Work backwards again in the problem. How long did Anita read books?** *1 hour*

5. **Work backwards on the clock. What time was it when Anita started reading?** *2 o'clock* **Write this time in the box where it belongs.**

6. **Work backwards again in the problem. How long did Anita play a game with her sister?** *1 hour*

7. **Work backwards on the clock. What time was it when Anita started playing a game with her sister?**

Solution: *1 o'clock*

LOOK BACK

Students should read or listen to the problem again and check their work. Encourage them to ask themselves, **Did I answer the question that was asked? Is my answer right?**

EXTENSION PROBLEM

On Saturday Mike played a game with his brother for 1 hour. Then he played with a friend for 1 hour. Then Mike and his mother rode their bicycles for 1 hour. They got home at 11 o'clock. What time was it when Mike started playing a game with his brother?

Solution: *8 o'clock*

TALK ABOUT IT

Have students talk with a partner or with a group about how they solved the Extension Problem. Students can share their different ways of thinking. Ask a question like, **What was the hardest thing about working backwards in this problem?**

WRITE YOUR OWN PROBLEM

Have students write similar problems of their own. Students can exchange problems and solve them.

PRACTICE

Similar Practice Problems: 78, 79, 80

When you give students a Practice Problem, ask questions such as, **Have you solved a problem like this before? What strategies helped you solve it?**

28   Jeff and his dad go to a basketball game. It takes them 1 hour to get there. They are at the game for 2 hours. It takes them 1 hour to get home. They get home at 7 o'clock. What time was it when Jeff and his dad left home for the game?

FIND OUT

- **What is the problem about?** Encourage students to restate the problem in their own words.

- **What do you have to find out to solve the problem?** *What time it was when Jeff and his dad left home for the game*

- **Find out what the problem tells you.**

  How long did it take them to get to the game? *1 hour*

  How long did they stay at the game? *2 hours*

  How long did it take them to get home? *1 hour*

  What time was it when they got home? *7 o'clock*

CHOOSE STRATEGIES

You can *Work Backwards* and *Use or Make a Picture or Diagram*. Use the picture of a clock face. Use both strategies to help you find out what time Jeff and his dad left home for the game.

 TEACHING TIP

Tell students they will be working backwards in two ways in this problem. They will begin at the end of the problem and work backwards through the problem. They will also work backwards on the clock. Have students practice going backwards on the clock. Ask them what time would be 1 hour before 2 o'clock.

SOLVE IT

Use the clock to find the times.

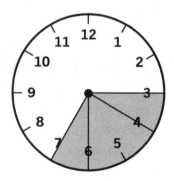

Write the times in the boxes.

Leave home	Get to the game	Leave the game	Get home
3 o'clock	**4** o'clock	**6** o'clock	**7** o'clock

1. **What time was it when Jeff and his dad got home?** *7 o'clock* **Where is this time given in the problem?** *At the end*

2. **Work backwards from the end of the problem. How long did it take Jeff and his dad to get home?** *1 hour*

3. **Work backwards on the clock. What time was it when they left the game?** *6 o'clock* **Write this time in the box where it belongs.**

4. **Work backwards again in the problem. How long did they stay at the game?** *2 hours*

5. **Work backwards on the clock. What time was it when Jeff and his dad got to the game?** *4 o'clock* **Write this time in the box where it belongs.**

6. **Work backwards again in the problem. How long did it take them to get to the game?** *1 hour*

7. **Work backwards on the clock. What time was it when Jeff and his dad left home for the game?**

Solution: *3 o'clock*

LOOK BACK

Students should read or listen to the problem again and check their work. Encourage them to ask themselves, **Did I answer the question that was asked? Is my answer right?**

EXTENSION PROBLEM

Nan and Brook went to visit their cousin. It took them 1 hour to get to their cousin's house. They stayed at their cousin's house for 2 hours. It took them 1 hour to get home. They got home at 2 o'clock. What time was it when Nan and Brook left home?

Solution: *10 o'clock*

TALK ABOUT IT

Have students talk with a partner or with a group about how they solved the Extension Problem. Students can share their different ways of thinking. Ask a question like, **What was hardest about working backwards?**

WRITE YOUR OWN PROBLEM

Have students write similar problems of their own. Students can exchange problems and solve them.

PRACTICE

Similar Practice Problems: 78, 79, 80

When you give students a Practice Problem, ask questions such as, **Have you solved a problem like this before? What strategies helped you solve it?**

Use or Look for a Pattern
Use or Make a Table

29 Professor Doogle found some dinosaur bones! On the first day he found 4 dinosaur bones. On the second day he found 8 bones. On the third day he found 12 bones. He sees a pattern in the numbers of bones. He keeps finding more bones each day in the same way. How many bones will the professor find on the fifth day?

FIND OUT

- **What is the problem about?** Encourage students to restate the problem in their own words.

- **What do you have to find out to solve the problem?** *How many bones the professor will find on the fifth day*

- **Find out what the problem tells you.**

  What did Professor Doogle find? *Dinosaur bones*

  How many bones did he find on the first day? *4 bones*

  How many bones did he find on the second day? *8 bones*

  How many bones did he find on the third day? *12 bones*

  What does the professor see in the numbers of bones? *A pattern*

  What do you know about how he keeps finding bones? *He finds more bones each day in the same way.*

CHOOSE STRATEGIES

You can *Use or Look for a Pattern* and *Use or Make a Table.* Use both strategies to help you find the number of bones Professor Doogle will find on the fifth day.

🍎 TEACHING TIP

Encourage students to talk about patterns. Have them describe what they think a pattern is. Talk about different kinds of number patterns. You can give examples such as 2, 4, 6, 8 or 12, 9, 6, 3. Point out that there can be patterns in numbers that get larger and patterns in numbers that get smaller.

SOLVE IT

Look at the table that has been started.

Day	Number of Bones
First	4
Second	**8**
Third	**12**
Fourth	**16**
Fifth	**20**

1. **How many bones did the professor find on the first day?** *4 bones* **Find this number in the table.**

2. **How many bones did he find on the second day?** *8 bones* **Write this number in the table.**

3. **How many bones did he find on the third day?** *12 bones* **Write this number in the table.**

4. **Look for a pattern in the numbers. What pattern do you see?** Students may describe the pattern in different ways. One possibility is shown. *The numbers are getting larger. Each day he finds 4 more bones than he found the day before.*

5. **Use the pattern. Finish the table. Keep writing numbers in the table until you find the number of bones on the fifth day.**

6. **How many bones will the professor find on the fifth day?**

   Solution: *20 bones*

LOOK BACK

Students should read or listen to the problem again and check their work. Encourage them to ask themselves, **Did I answer the question that was asked? Is my answer right?**

EXTENSION PROBLEM

How many bones will the professor find on the eighth day?

Solution: *32 bones*

TALK ABOUT IT

Have students talk with a partner or with a group about how they solved the Extension Problem. Students can share their different ways of thinking. Ask a question like, **How did you use the pattern?**

WRITE YOUR OWN PROBLEM

Have students write similar problems of their own. Students can exchange problems and solve them.

PRACTICE

Similar Practice Problems: 81, 82, 83, 100

When you give students a Practice Problem, ask questions such as, **Have you solved a problem like this before? What strategies helped you solve it?**

30 The three bears called Inspector Toad. On Monday there were 13 cookies in their cookie jar. On Tuesday there were 11 cookies in the jar. On Wednesday there were 9 cookies in the jar. Inspector Toad sees a pattern. The cookies keep disappearing each day in the same way. When will the cookie jar be empty?

FIND OUT

- **What is the problem about?** Encourage students to restate the problem in their own words.

- **What do you have to find out to solve the problem?** *When the cookie jar will be empty*

- **Find out what the problem tells you.**

   How many cookies did the bears have on Monday? *13 cookies*

   How many cookies did they have on Tuesday? *11 cookies*

   How many cookies did they have on Wednesday? *9 cookies*

   What does Inspector Toad notice? *A pattern*

   What do you know about how the cookies keep disappearing? *They disappear each day in the same way.*

CHOOSE STRATEGIES

You can *Use or Look for a Pattern* and *Use or Make a Table*. Use both strategies to help you find out when the jar will be empty.

 TEACHING TIP

Encourage students to talk about patterns. Have them describe what they think a pattern is. Talk about different kinds of number patterns. You can give examples such as 2, 4, 6, 8 or 16, 12, 8, 4. Point out that there can be patterns in numbers that get larger and patterns in numbers that get smaller.

SOLVE IT

Look at the table that has been started.

Day	Number of Cookies
Monday	13
Tuesday	11
Wednesday	9
Thursday	7
Friday	5
Saturday	3
Sunday	1
Monday	0

1. **How many cookies were in the jar on Monday?**
   13 cookies **Find this number in the table.**

2. **How many cookies were in the jar on Tuesday?**
   11 cookies **Write this number in the table.**

3. **How many cookies were in the jar on Wednesday?**
   9 cookies **Write this number in the table.**

4. **Look for a pattern in the numbers. What pattern do you see?** Students may describe the pattern in different ways. One possibility is shown.
   The numbers are getting smaller each time by 2.

5. **Use the pattern. Finish the table. Keep writing numbers in the table until the jar is empty.**

6. **When will the cookie jar be empty?**

   Solution: *Monday*

LOOK BACK

Students should read or listen to the problem again and check their work. Encourage them to ask themselves, **Did I answer the question that was asked? Is my answer right?**

EXTENSION PROBLEM

The next month the bears call Inspector Toad back. Their cookies are disappearing again! On Monday they had 16 cookies in their jar. On Tuesday there were 13 cookies in the jar. On Wednesday there were 10 cookies in the jar. The cookies keep disappearing each day in the same way. When will their jar be empty?

Solution: *Sunday*

TALK ABOUT IT

Have students talk with a partner or with a group about how they solved the Extension Problem. Students can share their different ways of thinking. Ask a question like, **How did you find the pattern?**

WRITE YOUR OWN PROBLEM

Have students write similar problems of their own. Students can exchange problems and solve them.

PRACTICE

Similar Practice Problems: 81, 82, 83, 100

When you give students a Practice Problem, ask questions such as, **Have you solved a problem like this before? What strategies helped you solve it?**

Use Logical Reasoning
Act Out or Use Objects

 Each pair needs: Pattern Blocks (1 each of square, triangle, hexagon, and trapezoid)

31 Noah and Vicky are playing a game called Mystery Shape. Noah hides a shape. He gives these clues about his mystery shape.

- It has more than 3 sides.
- All of its sides are the same length.
- It does not have square corners.

Which of these shapes is Noah's mystery shape?

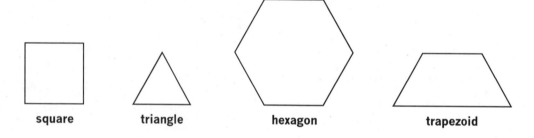

square   triangle   hexagon   trapezoid

FIND OUT

- **What is the problem about?** Encourage students to restate the problem in their own words.

- **What do you have to find out to solve the problem?** *Which shape is Noah's mystery shape*

- **Find out what the problem tells you.**

  What clues does Noah give about his mystery shape? *(1) It has more than 3 sides. (2) All of its sides are the same length. (3) It does not have square corners.*

  What shapes are shown? *A square, a triangle, a hexagon, and a trapezoid*

 TEACHING TIP

Review the names of the shapes with students. They may also need to review the meanings of the words *side*, *corner (angle)*, and *square corner*. Then ask them which of the shapes has square corners. *(the square)*

CHOOSE STRATEGIES

You can *Use Logical Reasoning* and *Act Out or Use Objects*. Use the clues and think about them this way: If you know what shapes fit the clue, then you know which shapes do not fit the clue.

SOLVE IT

Use Pattern Blocks to show the shapes. Use the clues to find the mystery shape.

1. **What is the first clue?** *It has more than 3 sides.*
If the shape has more than 3 sides, then which shapes could it be? *The square, the hexagon, and the trapezoid* **Are there any shapes that it could NOT be?** *Yes, the triangle* **Write NO under that shape.**

2. **What is the second clue?** *All of its sides are the same length.* **If all of its sides are the same length, then which shapes could it be?** *The square and the hexagon* **Are there any shapes that it could NOT be?** *Yes, the trapezoid* **Write NO under that shape.**

3. **What is the third clue?** *It does not have square corners.* **If the shape does not have square corners, are there any shapes that it could NOT be?** *Yes, the square* **Write NO under that shape.**

4. **Which shape is Noah's mystery shape?**

Solution: *The hexagon*

LOOK BACK

Students should read or listen to the problem again and check their work. Encourage them to ask themselves, **Did I answer the question that was asked in the problem? Is my answer right?**

EXTENSION PROBLEM

Now Vicky hides one of the shapes. She gives these clues about her mystery shape:
- **It has fewer than 6 corners.**
- **All of its sides are the same length.**
- **It is a rectangle.**
Which of the shapes is Vicky's mystery shape?

Solution: *The square*

TALK ABOUT IT

Have students talk with a partner or small group about how they solved the Extension Problem. Students can share their different ways of thinking. Ask a question such as, **How did you use the Pattern Blocks to help you solve the problem?**

WRITE YOUR OWN PROBLEM

Have students write similar problems of their own. Students can then exchange problems and solve them.

PRACTICE

Similar Practice Problems: 84, 85, 86

When you give students a Practice Problem, ask questions such as, **Have you solved a problem like this before? What strategies helped you solve it?**

Use Logical Reasoning
Act Out or Use Objects

 Each pair needs: Pattern Blocks (1 trapezoid, 1 blue parallelogram, 2 triangles, 2 squares)

32 Erik and Kristin are playing a game called Take a Shape. Kristin draws a card that gives three clues.

> **You may take this shape:**
> • **It has no square corners.**
> • **It has more than 3 sides.**
> • **Its sides are not all the same length.**

Which shape may Kristin take?

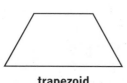      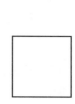   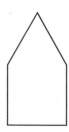

trapezoid   triangle   parallelogram   square   pentagon

FIND OUT

• **What is the problem about?** Encourage students to restate the problem in their own words.

• **What do you have to find out to solve the problem?** *Which shape Kristin may take*

• **Find out what the problem tells you.**

What are the clues that tell Kristin which shape she may take? *(1) The shape has no square corners. (2) It has more than 3 sides. (3) Its sides are not all the same length.*

What shapes are shown? *A trapezoid, a triangle, a parallelogram, a square, and a pentagon*

 TEACHING TIP

Review the names of the shapes with students. They may also need to review the meanings of the words *side, corner (angle),* and *square corner.* Then ask them to look at the shapes and find one that has square corners. *(the square or the pentagon)*

CHOOSE STRATEGIES

You can *Use Logical Reasoning* and *Act Out or Use Objects.* Use the clues and think about them this way: If you know what shapes fit the clue, then you know which shapes do not fit the clue.

SOLVE IT

Use Pattern Blocks to show the shapes. Use the clues to find the shape.

As students select the Pattern Blocks that match the shapes shown, help them see that they can put 2 blocks together (a square and a triangle) to make the pentagon.

1. **What is the first clue?** *It has no square corners.* **If the shape has no square corners, then which shapes could it be?** *The trapezoid, the triangle, and the parallelogram* **Are there any shapes that it could NOT be?** *Yes, the square and the pentagon* **Write NO under those shapes.**

2. **What is the second clue?** *It has more than 3 sides.* **If the shape has more than 3 sides, then which shapes could it be?** *The trapezoid and the parallelogram* **Are there any shapes that it could NOT be?** *Yes, the triangle* **Write NO under that shape.**

3. **What is the third clue?** *Its sides are not all the same length.* **If its sides are not all the same length, are there any shapes that it could NOT be?** *Yes, the parallelogram* **Write NO under that shape.**

4. **Which shape may Kristin take?**

   Solution: *The trapezoid*

LOOK BACK

Students should read or listen to the problem again and check their work. Encourage them to ask themselves, **Did I answer the question that was asked in the problem? Is my answer right?**

EXTENSION PROBLEM

Now Erik draws a card. It gives these clues:
- **The shape has sides that are all the same length.**
- **Its corners are not all the same size.**
- **The shape has more sides than the trapezoid.**

Which shape may Erik take?

Solution: *The pentagon*

TALK ABOUT IT

Have students talk with a partner or small group about how they solved the Extension Problem. Students can share their different ways of thinking. Ask questions like, **How did you use "if ... then" thinking to help you solve the problem? How did the Pattern Blocks help you?**

WRITE YOUR OWN PROBLEM

Have students write similar problems of their own. Students can then exchange problems and solve them.

PRACTICE

Similar Practice Problems: 84, 85, 86

When you give students a Practice Problem, ask questions such as, **Have you solved a problem like this before? What strategies helped you solve it?**

33 Some friends played together after school. There were 6 children who rode bikes. There were 4 children who rode scooters. Two of the children rode both bikes AND scooters. How many friends in all played together after school?

FIND OUT

- **What is the problem about?** Encourage students to restate the problem in their own words.

- **What do you have to find out to solve the problem?** *How many friends in all played together after school*

- **Find out what the problem tells you.**

   How many children rode bikes? *6 children*

   How many children rode scooters? *4 children*

   How many children rode both bikes AND scooters? *2 children*

CHOOSE STRATEGIES

You can *Use or Make a Picture or Diagram* and *Use Logical Reasoning*. Use both strategies to help you find out how many friends in all played together after school.

SOLVE IT

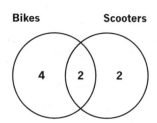

Look at the circle diagram. Why is the number 2 in the overlapping part of the circles? *Because this number of children rode both bikes and scooters*

 TEACHING TIP

Talk with students about the circle diagram (also called a Venn diagram). Point out that there are two circles, one for bikes and one for scooters. Then point out the space where the two circles overlap. Help students see that this space is part of both the bike circle and the scooter circle. Ask them what number goes in this space. *(The number of children who rode both bikes **and** scooters)* Be sure students understand that the rest of the space in each circle is for the number of children who **only** rode bikes (left circle), and the number who **only** rode scooters (right circle).

1. **How many children rode bikes?** *6 children*
Explain that this number includes the number of children who rode both bikes and scooters, as well as those who only rode bikes.

2. **How can you find the number of children who ONLY rode bikes?** *Subtract 2 from the 6 children in all who rode bikes.*

3. **How many children ONLY rode bikes?** *4 children*
Write this number where it belongs.

4. **How many children rode scooters?** *4 children*
Explain that this number includes the number of children who rode both bikes and scooters, as well as the ones who only rode scooters.

5. **Find the number of children who ONLY rode scooters. Write this number where it belongs.**

6. **How can you find out how many friends in all were playing together?** *Add the numbers in all 3 parts of the circles.*

7. **How many friends in all played together after school?**

Solution: *8 friends*

LOOK BACK

Students should read or listen to the problem again and check their work. Encourage them to ask themselves, **Did I answer the question that was asked? Is my answer right?**

EXTENSION PROBLEM

Another group of friends played together the next day. There were 7 children who rode bikes. There were 9 children who rode scooters. Four of the children rode both bikes AND scooters. How many friends in all played together the next day?

Solution: *12 friends*

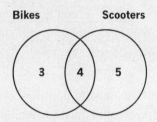

TALK ABOUT IT

Have students talk with a partner or with a group about how they solved the Extension Problem. Students can share their different ways of thinking. Ask a question like, **How does using a circle diagram help you solve this problem?**

WRITE YOUR OWN PROBLEM

Have students write similar problems of their own. Students can exchange problems and solve them.

PRACTICE

Similar Practice Problems: 87, 88, 89, 98

When you give students a Practice Problem, ask questions such as, **Have you solved a problem like this before? What strategies helped you solve it?**

Use or Make a Picture or Diagram
Use Logical Reasoning

34 The first graders played outside after lunch. There were 8 children who played tag. There were 10 children who played kickball. Three of the children played both tag AND kickball. How many children in all were playing these games?

FIND OUT

- **What is the problem about?** Encourage students to restate the problem in their own words.

- **What do you have to find out to solve the problem?** *How many children in all were playing these games*

- **Find out what the problem tells you.**

  How many children played tag? *8 children*

  How many children played kickball? *10 children*

  How many children played both tag AND kickball? *3 children*

CHOOSE STRATEGIES

You can *Use or Make a Picture or Diagram* and *Use Logical Reasoning*. Use both strategies to help you find out how many children in all played tag and kickball.

SOLVE IT

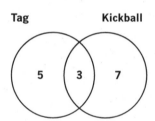

Look at the circle diagram. Why is the number 3 in the overlapping part of the circles? *Because this number of children played both tag and kickball*

 TEACHING TIP

Talk with students about the circle diagram (also called a Venn diagram): Point out that there are two circles, one for tag and one for kickball. Then point out the space where the two circles overlap. Help students see that this space is part of both the tag circle and the kickball circle. Ask what number would go in this space. *(The number of children who played both tag **and** kickball)* Be sure students understand that the space in the rest of each circle is for the number of children who **only** played tag (left circle) and for the number who **only** played kickball (right circle).

1. **How many children played tag?** *8 children*
   Explain that this number includes the number of children who played both tag and kickball, as well as those who only played tag.

2. **How can you find the number of children who ONLY played tag?** *Subtract 3 from the 8 children in all who played tag.*

3. **How many children ONLY played tag?** *5 children*
   Write this number where it belongs.

4. **How many children played kickball?** *10 children*
   Explain that this number includes the number of children who played both tag and kickball, as well as those who only played kickball.

5. **Find the number of children who ONLY played kickball. Write this number where it belongs.**

6. **How can you find out how many children in all played tag and kickball?** *Add the numbers in all 3 parts of the circles.*

7. **How many children in all were playing these games?**
   Solution: *15 children*

LOOK BACK

Students should read or listen to the problem again and check their work. Encourage them to ask themselves, **Did I answer the question that was asked? Is my answer right?**

EXTENSION PROBLEM

The next day there were 9 children who played tag. There were 12 children who played kickball. Five of those children played both tag AND kickball. How many children in all were playing these games?

Solution: *16 children*

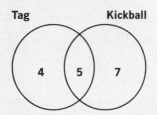

TALK ABOUT IT

Have students talk with a partner or with a group about how they solved the Extension Problem. Students can share their different ways of thinking. Ask a question like, **How does using a circle diagram help you solve this problem?**

WRITE YOUR OWN PROBLEM

Have students write similar problems of their own. Students can exchange problems and solve them.

PRACTICE

Similar Practice Problems: 87, 88, 89, 98

When you give students a Practice Problem, ask questions such as, **Have you solved a problem like this before? What strategies helped you solve it?**

Brainstorm
Act Out or Use Objects

 Each pair needs: 3 pennies,
3 nickels, and 3 dimes

35 Can you solve this puzzle? Take 3 pennies, 3 nickels,
and 3 dimes. Put the coins in three rows, like this:

How can you move only 2 coins and get the same
amount of money in every row and every column?

FIND OUT

- **What is the problem about?** Encourage students
  to restate the problem in their own words.

- **What do you have to find out to solve the problem?**
  *How to move only 2 coins so that you have the
  same amount of money in every row and column*

- **Find out what the problem tells you.**

  What coins do you take? *3 pennies, 3 nickels,
  3 dimes*

  What do you do with the coins? *Put the coins in
  three rows.*

  What does the puzzle ask you to do? *Move only
  2 coins so that there is the same amount of money
  in every row and column.*

 TEACHING TIP

To review the meaning of *rows* and *columns*, have students
name the coins in the top row (dime, nickel, nickel). Then
have students name the coins in the first column, reading
from the top to the bottom (dime, penny, penny).

CHOOSE STRATEGIES

**You can *Brainstorm* and *Act Out or Use Objects*.
Brainstorming is a special kind of thinking.
Sometimes it means thinking about something
in unusual ways. It means trying out many ideas
until the answer pops into your mind. Use play
money to show the coins.**

SOLVE IT

Put the coins on the grid.

1. **How many rows of coins are there?** *3 rows*

2. **What is the value of the coins in the first row?**
20 cents

3. **What is the value of the coins in the second row?**
16 cents

4. **What is the value of the coins in the third row?**
12 cents

5. **Now find the value of the coins in each column.**

6. **Try different ways of moving just 2 coins to solve the puzzle.**

7. **How can you move only 2 coins and get the same amount of money in every row and every column?**

Solution: *Switch the penny at the bottom of the first column with the nickel on the top of the third column.*

LOOK BACK

Students should read or listen to the problem again and check their work. Encourage them to ask themselves, **Did I answer the question that was asked? Is my answer right?**

EXTENSION PROBLEM

Here's another coin puzzle to solve. Put 3 dimes, 3 nickels, and 3 pennies in a triangle, like this:

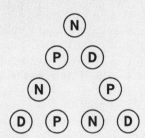

Now move only 2 coins so that you show 26 cents on every side of the triangle.

How can you solve this puzzle?

Solution: *Switch the nickel at the top of the triangle with the dime next to it on the right side of the triangle.*

TALK ABOUT IT

Have students talk with a partner or with a group about how they solved the Extra Challenge. Students can share their different ways of thinking. Ask questions like, **What ideas did you have when you brainstormed? How did you get started?**

WRITE YOUR OWN PROBLEM

Have students write similar problems of their own. Students can exchange problems and solve them.

PRACTICE

Similar Practice Problems: 90, 91, 92

When you give students a Practice Problem, ask questions such as, **Have you solved a problem like this before? What strategies helped you solve it?**

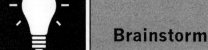
36 Hong gave this code puzzle to her sister Lin:

2 E   2 E   1 N   1 M

Lin solved the puzzle. What does the code puzzle mean?

FIND OUT

- **What is the problem about?** Encourage students to restate the problem in their own words.

- **What do you have to find out to solve the problem?** *Find out what the code puzzle means*

- **Find out what the problem tells you.**

  What are the numbers and letters in the code? *2 E, 2 E, 1 N, 1 M*

CHOOSE STRATEGIES

You can *Brainstorm* to help you solve this kind of problem. Brainstorming is a special kind of thinking. Sometimes it means thinking about something in unusual ways. It means trying out many ideas until the answer pops into your mind.

SOLVE IT

Encourage all students to offer their ideas. The greater the variety of ideas the better! The answers below include samples of the kinds of ideas students might have.

1. **What is shown in the code?** *Numbers and letters*

2. **What numbers are the same?** *Two 2s and two 1s*

3. **What letters are the same?** *Two Es*

4. **What could the E be a code for?** *Eagle, eel, egg*

5. **If the 2 goes with the E, what could 2 E be?** *2 elephants*

6. **Could the E be a code for something you have?** Again, encourage any and all ideas.

7. **What do you have two of that begins with E?** *Eyes*

8. **Do you have anything else that begins with E?** *Ears*

9. **Now think about the N and M. What number goes with these letters?** *The number 1* **If these all go together, what do you have that might go with ears and eyes?** *Nose and mouth*

10. **What does the code puzzle mean?**

    Solution: *2 eyes, 2 ears, 1 nose, 1 mouth*

LOOK BACK

Students should read or listen to the problem again and check their work. Encourage them to ask themselves, **Did I answer the question that was asked? Is my answer right?**

EXTENSION PROBLEM

Then Lin gave Hong a puzzle:

2 H   2 F   10 F   10 T

Hong solved the puzzle. What does the code puzzle mean?

Solution: *2 hands, 2 feet, 10 fingers, 10 toes*

TALK ABOUT IT

Have students talk with a partner or with a group about how they solved the Extra Challenge. Students can share their different ways of thinking. Ask questions like, **What ideas did you have when you brainstormed? How did you get started?**

WRITE YOUR OWN PROBLEM

Have students write similar problems of their own. Students can exchange problems and solve them.

PRACTICE

Similar Practice Problems: 90, 91, 92

When you give students a Practice Problem, ask questions such as, **Have you solved a problem like this before? What strategies helped you solve it?**

Use Logical Reasoning
Act Out or Use Objects

 Each pair needs:
8 nickels

37   In the Wild-at-the-Zoo Shop,

   costs the same as

If 1  costs 5 cents, how much does 1 cost?

FIND OUT

- **What is the problem about?** Encourage students to restate the problem in their own words.

- **What do you have to find out to solve the problem?**
  How much 1 lion costs

- **Find out what the problem tells you.**

  What do you know about how much 1 lion costs?
  It costs the same as 3 lizards.

  How much does 1 lizard cost? *5 cents*

CHOOSE STRATEGIES

You can *Use Logical Reasoning* and *Act Out or Use Objects.* Use a special kind of thinking. IF you know how much one kind of animal costs, THEN you can figure out how much the other kind costs. Use play money to show the costs.

SOLVE IT

1. **What do you know about how much 1 lion costs?**
   It costs the same as 3 lizards.

2. **How much does 1 lizard cost?** *5 cents* **Show that with a coin.**

   How can you find out how much 3 lizards cost?
   Start with 5 cents for 1 lizard, then add 5 cents for every other lizard. **Show that with your coins.**

3. **How much do 3 lizards cost?** *15 cents*

4. **Then how much does 1 lion cost?**

   Solution: *15 cents*

LOOK BACK

Students should read or listen to the problem again and check their work. Encourage them to ask themselves, **Did I answer the question that was asked in the problem? Is my answer right?**

EXTENSION PROBLEM

In the Wild-at-the-Zoo Shop, 1 zebra costs the same as 5 wildcats. If 1 wildcat costs 5 cents, how much does 1 zebra cost?

Solution: *25 cents*

TALK ABOUT IT

Have students talk with a partner or small group about how they solved the Extension Problem. Students can share their different ways of thinking. Ask questions like, **How did you use the play money to help you solve the problem? How did "if ... then" thinking help you?**

WRITE YOUR OWN PROBLEM

Have students write similar problems of their own. Students can then exchange problems and solve them.

PRACTICE

Similar Practice Problems: 93, 94, 95

When you give students a Practice Problem, ask questions such as, **Have you solved a problem like this before? What strategies helped you solve it?**

Use Logical Reasoning
Act Out or Use Objects

 Each pair needs:
10 dimes

38 In the Seaside Gift Shop,

 costs the same as

If 1  costs 10 cents, how much does 1  cost?

FIND OUT

- **What is the problem about?** Encourage students to restate the problem in their own words.

- **What do you have to find out to solve the problem?** *How much 1 dolphin costs*

- **Find out what the problem tells you.**

  What do you know about how much 1 dolphin costs? *It costs the same as 4 turtles.*

  How much does 1 turtle cost? *10 cents*

CHOOSE STRATEGIES

You can *Use Logical Reasoning* and *Act Out or Use Objects*. Use a special kind of thinking. IF you know how much one kind of sea animal costs, THEN you can figure out how much the other kind costs. Use play money to show the costs.

SOLVE IT

1. **What do you know about how much 1 dolphin costs?** *It costs the same as 4 turtles.*

2. **How much does 1 turtle cost?** *10 cents* **Show that with a coin.**

   How can you find out how much 4 turtles cost? *Start with 10 cents for 1 turtle, then add 10 cents for every other turtle.* **Show that with your coins.**

3. **How much do 4 turtles cost?** *40 cents*
4. **Then how much does 1 dolphin cost?**

   Solution: *40 cents*

LOOK BACK

Students should read or listen to the problem again and check their work. Encourage them to ask themselves, **Did I answer the question that was asked in the problem? Is my answer right?**

Problem 38 : Algebra, Number and Operations

EXTENSION PROBLEM

In the Seaside Gift Shop, 1 octopus costs the same as 5 snails. If 1 snail costs 10 cents, how much does 1 octopus cost?

Solution: *50 cents*

TALK ABOUT IT

Have students talk with a partner or small group about how they solved the Extension Problem. Students can share their different ways of thinking. Ask questions like, **How did you use the coins to help you solve the problem? How did "if ... then" thinking help you?**

WRITE YOUR OWN PROBLEM

Have students write similar problems of their own. Students can then exchange problems and solve them.

PRACTICE

Similar Practice Problems: 93, 94, 95

When you give students a Practice Problem, ask questions such as, **Have you solved a problem like this before? What strategies helped you solve it?**

Name _____

There were 15 pirates on the shore. They got into 3 boats. Now there are 12 pirates in all in the first boat and second boat. There are 8 pirates in all in the second boat and third boat. How many pirates are in each boat?

First

Second

Third

FIND OUT

What do you have to find out to solve the problem?

CHOOSE STRATEGIES

Ring the strategies you use.

SOLVE IT

Show your work.

LOOK BACK

Read the problem again. Check your work.

Name _____

Melanie put 17 Fruit Dinos into 3 bags. Now there are 8 dinos in all in the first bag and second bag. There are 16 dinos in all in the second bag and third bag. How many dinos are in each bag?

First

Second

Third

FIND OUT
What do you have to find out to solve the problem?

CHOOSE STRATEGIES
Ring the strategies you use.

SOLVE IT
Show your work.

LOOK BACK
Read the problem again. Check your work.

Name _____

Jerry had 18 pennies. He made 3 piles of pennies. Now there are 16 pennies in all in the first pile and second pile. There are 7 pennies in all in the second pile and third pile. How many pennies are in each pile?

FIND OUT

What do you have to find out to solve the problem?

CHOOSE STRATEGIES

Ring the strategies you use.

SOLVE IT

Show your work.

LOOK BACK

Read the problem again. Check your work.

42

Name _____

Tessa and Eli are drawing red frogs on paper hats for a party. They draw 4 frogs on each hat. Tessa and Eli draw 28 frogs in all. How many paper hats do they have?

FIND OUT

What do you have to find out to solve the problem?

CHOOSE STRATEGIES

Ring the strategies you use.

SOLVE IT

Show your work.

LOOK BACK

Read the problem again. Check your work.

Name _____

Rosa and her friends are going to ride
on the Blue Monster Train. Five children
get into each car. When 30 children
are in the cars, the train begins to move.
How many cars are the children in?

FIND OUT

What do you have to find out to solve the problem?

CHOOSE STRATEGIES

Ring the strategies you use.

SOLVE IT

Show your work.

LOOK BACK

Read the problem again. Check your work.

Name _____

Elisa and Rick are playing Hippo Hop. They get 6 points every time their game marker lands on a hippo. Rick has 36 points already. How many times has his marker landed on a hippo?

FIND OUT
What do you have to find out to solve the problem?

CHOOSE STRATEGIES
Ring the strategies you use.

SOLVE IT
Show your work.

LOOK BACK
Read the problem again. Check your work.

Name _____

Abel's class is collecting teddy bears. The bears are for children who lost their homes in a fire. The first day the class got 2 bears. The second day they got 4 bears. The third day they got 6 bears. Every day they got 2 more bears than they got the day before. How many bears did they get on the sixth day?

FIND OUT
What do you have to find out to solve the problem?

CHOOSE STRATEGIES
Ring the strategies you use.

SOLVE IT
Show your work.

LOOK BACK
Read the problem again. Check your work.

46

Name _____

Rebecca Rabbit and her brother Peter hopped into the flower garden. They ate 3 flowers the first day. They ate 6 flowers the second day. They ate 9 flowers the third day. Each day they ate 3 more flowers than they ate the day before. How many flowers did they eat on the fifth day?

FIND OUT
What do you have to find out to solve the problem?

CHOOSE STRATEGIES
Ring the strategies you use.

SOLVE IT
Show your work.

LOOK BACK
Read the problem again. Check your work.

Name _____

Carmen and her cousins love to hear their
Granddaddy tell stories. On Sunday he
tells 3 stories. On Monday he tells 5 stories.
On Tuesday he tells 7 stories. Every day
he tells 2 more stories than he told the
day before. How many stories will he tell
on Saturday?

FIND OUT

What do you have to find out to solve the problem?

CHOOSE STRATEGIES

Ring the strategies you use.

SOLVE IT

Show your work.

LOOK BACK

Read the problem again. Check your work.

48

Name _____

Kayla has 15 tiny monsters
in all. She has 5 more purple
monsters than orange monsters.
How many monsters of each
color does she have?

FIND OUT
What do you have to find out to solve the problem?

CHOOSE STRATEGIES
Ring the strategies you use.

SOLVE IT
Show your work.

LOOK BACK
Read the problem again. Check your work.

Name _____

Becky takes 16 coins out of her pocket.
There are 6 more pennies than dimes.
How many pennies and how many dimes
does Becky have?

FIND OUT
What do you have to find out to solve the problem?

CHOOSE STRATEGIES
Ring the strategies you use.

SOLVE IT
Show your work.

LOOK BACK
Read the problem again. Check your work.

50

Name _____

Tracy loves T-shirts! She has 11 in all. She has the same number of red shirts as blue shirts. She has 2 fewer red shirts than pink shirts. How many shirts of each color does she have?

FIND OUT

What do you have to find out to solve the problem?

CHOOSE STRATEGIES

Ring the strategies you use.

SOLVE IT

Show your work.

LOOK BACK

Read the problem again. Check your work.

Name _____

The girls are at Nature Camp. They are looking for lizards. They see 3 lizards on the first day. They see 6 lizards on the second day. They see 9 lizards on the third day. They keep seeing more lizards in the same way. How many lizards will the girls see on the sixth day?

FIND OUT
What do you have to find out to solve the problem?

CHOOSE STRATEGIES
Ring the strategies you use.

SOLVE IT
Show your work.

LOOK BACK
Read the problem again. Check your work.

Name _____

The Three Little Pigs went to market to buy
bricks. On Monday they bought 3 bricks.
On Tuesday they bought 5 bricks. On Wednesday
they bought 7 bricks. They kept buying bricks
in the same way. How many bricks did they buy
on Saturday?

FIND OUT
What do you have to find out to solve the problem?

CHOOSE STRATEGIES
Ring the strategies you use.

SOLVE IT
Show your work.

LOOK BACK
Read the problem again. Check your work.

Name _____

On Monday, Farmer Fox counts 20 ears of corn in his garden. On Tuesday he counts 17 ears of corn in his garden. On Wednesday he counts 14 ears of corn in his garden. The ears of corn keep disappearing in the same way. How many ears of corn will Farmer Fox count in his garden on Saturday?

FIND OUT
What do you have to find out to solve the problem?

CHOOSE STRATEGIES
Ring the strategies you use.

SOLVE IT
Show your work.

LOOK BACK
Read the problem again. Check your work.

Name _____

It's time for lunch at Burgers to Go. Five cars are in line. There is 1 red, 1 yellow, 1 blue, and 2 white cars. The first car and the last car are the same color. The yellow car is between the red car and blue car. The blue car is next to last in line. What color is each car in the line?

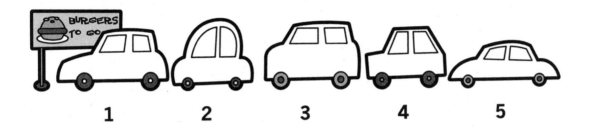

1 2 3 4 5

FIND OUT
What do you have to find out to solve the problem?

CHOOSE STRATEGIES
Ring the strategies you use.

SOLVE IT
Show your work.

LOOK BACK
Read the problem again. Check your work.

Name _____

A train with 5 cars rolls into the station. There are 2 red cars behind a yellow car. A green car is behind the 2 red cars. The blue car is not last in line. What color is each train car in the line, from first to last?

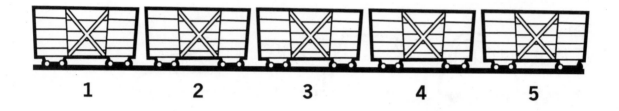

1   2   3   4   5

FIND OUT
What do you have to find out to solve the problem?

CHOOSE STRATEGIES
Ring the strategies you use.

SOLVE IT
Show your work.

LOOK BACK
Read the problem again. Check your work.

56

There are 6 flags on a pole.
A yellow flag is at the top. A red
flag is between 2 yellow flags.
A red flag is between a blue flag
and green flag. A blue flag is
not at the bottom. What color is
each flag on the pole, from top
to bottom?

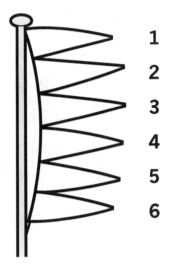

1
2
3
4
5
6

FIND OUT

What do you have to find out to solve the problem?

CHOOSE STRATEGIES

Ring the strategies you use.

SOLVE IT

Show your work.

LOOK BACK

Read the problem again. Check your work.

Name _____

Matt's father gave him 4 nickels, 4 dimes, and 4 pennies. Matt put 2 of the coins into each of his pockets. Each pocket has a different group of coins in it than every other pocket. What coins could be in each pocket?

FIND OUT
What do you have to find out to solve the problem?

CHOOSE STRATEGIES
Ring the strategies you use.

SOLVE IT
Show your work.

LOOK BACK
Read the problem again. Check your work.

Name _____

The elves had 3 yellow feathers, 3 green
feathers, 3 red feathers, and 3 blue feathers.
Each elf took 2 feathers and put them on
his cap. Each elf took a different group of
2 feathers than every other elf. What colors
are the feathers on each cap?

Hint: You may find more than one answer.

FIND OUT

What do you have to find out to solve the problem?

CHOOSE STRATEGIES

Ring the strategies you use.

SOLVE IT

Show your work.

LOOK BACK

Read the problem again. Check your work.

Name _____

In the Frog Marching Band, there
are 2 blue frogs, 2 yellow frogs,
2 red frogs, and 4 green frogs. The
frogs are marching in rows. There
are 2 frogs in each row. Each row is
different from every other row. What
colors are the frogs in each row?

Hint: You may find more than one
answer.

FIND OUT
What do you have to find out to solve the problem?

CHOOSE STRATEGIES
Ring the strategies you use.

SOLVE IT
Show your work.

LOOK BACK
Read the problem again. Check your work.

Permission is given to instructors to reproduce this page for classroom use with *Problem Solver II*. Copyright ©2004 Wright Group/McGraw-Hill.

60

Name _____

Marla has 2 flower pots.
She is putting 7 seeds
into the pots. She puts at
least 1 seed in each pot.
What are all the different
ways that she can put
the seeds in the pots?

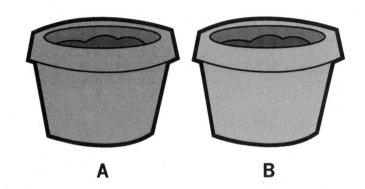

A   B

FIND OUT

What do you have to find out to solve the problem?

CHOOSE STRATEGIES

Ring the strategies you use.

SOLVE IT

Show your work.

LOOK BACK

Read the problem again. Check your work.

Name _____

Bob has 2 piggy banks and 8 dimes. He is putting the dimes into the piggy banks. He puts at least 1 dime into each bank. What are all the different ways that he can put the dimes into the banks?

A   **B**

FIND OUT

What do you have to find out to solve the problem?

CHOOSE STRATEGIES

Ring the strategies you use.

SOLVE IT

Show your work.

LOOK BACK

Read the problem again. Check your work.

62

Name _____

Nine friends are going on the **Wave Ride**. The friends are getting into 2 boats. At least 1 friend gets into each boat. What are all the different ways that the friends could be in the boats?

A   **B**

FIND OUT

What do you have to find out to solve the problem?

CHOOSE STRATEGIES

Ring the strategies you use.

SOLVE IT

Show your work.

LOOK BACK

Read the problem again. Check your work.

Name _____

Jodi and Lisette are playing a board game. Lisette's game marker is on space 1 of the track. She moves her marker forward 4 spaces. Next, she moves her marker back 1 space. Next, she moves her marker forward 11 spaces. Jodi's game marker is on space 13. Who is ahead, Jodi or Lisette?

FIND OUT
What do you have to find out to solve the problem?

CHOOSE STRATEGIES
Ring the strategies you use.

SOLVE IT
Show your work.

1															

LOOK BACK
Read the problem again. Check your work.

Name _____

Diane dropped some coins on the steps. Now she is trying to find them all. She starts at the bottom of the steps. She goes up 3 steps and finds a dime. She goes up 3 more steps and finds a nickel. Then she goes down 2 steps and finds a quarter. Then she goes up 7 steps and finds a penny. What step does Diane find the penny on?

FIND OUT
What do you have to find out to solve the problem?

CHOOSE STRATEGIES
Ring the strategies you use.

SOLVE IT
Show your work.

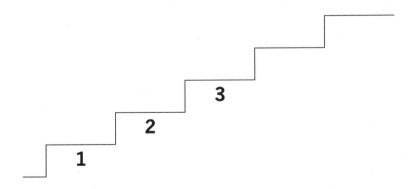

LOOK BACK
Read the problem again. Check your work.

65

Jon is looking for his lost mitt. It is in a
locker. He starts at the first locker in the
row. Then he goes to the right 5 lockers.
Next he goes back 4 lockers. Then he
goes right 8 lockers. Then he goes back
3 lockers. There is his mitt! What locker
is his mitt in?

FIND OUT
What do you have to find out to solve the problem?

CHOOSE STRATEGIES
Ring the strategies you use.

SOLVE IT
Show your work.

1	2	3	4	5	6	7	

LOOK BACK
Read the problem again. Check your work.

Name _____

Kim bought crickets for her
iguana. She paid more than
23 cents and less than 33 cents.
She paid with 3 pennies and
3 other coins. How much did
Kim pay?

FIND OUT

What do you have to find out to solve the problem?

CHOOSE STRATEGIES

Ring the strategies you use.

SOLVE IT

Show your work.

LOOK BACK

Read the problem again. Check your work.

Name _____

Eve and Hunter bought some chewy stars. They paid less than 40 cents and more than 30 cents. They paid with 3 nickels and 3 other coins. How much did Eve and Hunter pay?

FIND OUT
What do you have to find out to solve the problem?

CHOOSE STRATEGIES
Ring the strategies you use.

SOLVE IT
Show your work.

LOOK BACK
Read the problem again. Check your work.

Name _____

Kenya and Marcy bought some charms.
Together they paid more than 36 cents
and less than 45 cents. They paid with
a quarter and 2 other coins. How much
did Kenya and Marcy pay?

FIND OUT

What do you have to find out to solve the problem?

CHOOSE STRATEGIES

Ring the strategies you use.

SOLVE IT

Show your work.

LOOK BACK

Read the problem again. Check your work.

Name _____

Billy Bear really likes honey!
He also likes fish, and he loves
berry pie. Billy Bear starts
at the tree. He takes this path
to find some lunch:

- 1 step to the right
- 2 steps up
- 3 steps to the right
- 4 steps up

Billy eats any food on his
path. What food does Billy miss?

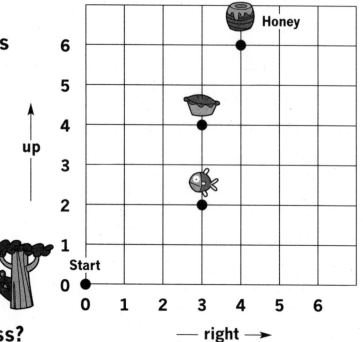

FIND OUT

What do you have to find out to solve the problem?

CHOOSE STRATEGIES

Ring the strategies you use.

SOLVE IT

Show your work.

LOOK BACK

Read the problem again. Check your work.

Name _____

Franny Frog starts at the log. She wants some flies! She takes this path:

- 2 hops to the right
- 5 hops up
- 3 hops to the right
- 1 hop up

Franny eats every fly on her path. How many flies does Franny eat?

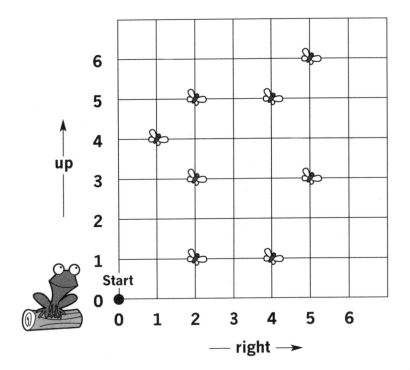

FIND OUT
What do you have to find out to solve the problem?

CHOOSE STRATEGIES
Ring the strategies you use.

SOLVE IT
Show your work.

LOOK BACK
Read the problem again. Check your work.

Name _____

The mother kangaroo starts
out at the bush. She is looking
for Joey, her young kangaroo.
She takes this path:

- **4 hops to the right**
- **4 hops up**
- **1 hop to the right**
- **1 hop up**
- **2 hops to the right**

There's Joey! What
different path could the mother
take? Use the same number of hops.

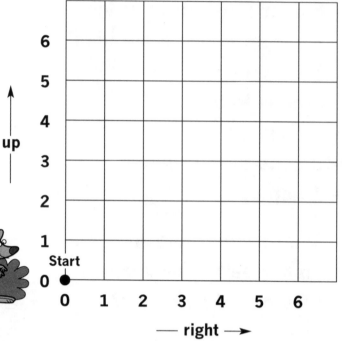

FIND OUT

What do you have to find out to solve the problem?

CHOOSE STRATEGIES

Ring the strategies you use.

SOLVE IT

Show your work.

LOOK BACK

Read the problem again. Check your work.

Name _____

Our Favorite Rides at Magic Land

Boat	☺☺☺☺☺☺☺😀😀😀😀😀😀
Roller coaster	☺☺☺☺☺☺😀😀😀
Ferris wheel	☺☺☺☺😀😀😀😀
Merry-go-round	☺☺😀😀

😀 is 1 boy

☺ is 1 girl

The first grade voted for their favorite rides at Magic Land. Look at the graph. Use the clues to find out how Melody, Julie, and Andre voted.

- **The same number of girls as boys liked Melody's favorite ride.**
- **2 more girls than boys liked Julie's favorite ride.**
- **Andre's ride has 5 more votes than the Ferris wheel.**

What rides did Melody, Julie, and Andre vote for?

FIND OUT

What do you have to find out to solve the problem?

CHOOSE STRATEGIES

Ring the strategies you use.

SOLVE IT

Show your work.

LOOK BACK

Read the problem again. Check your work.

Name _____

Things We Like to Do at the Park

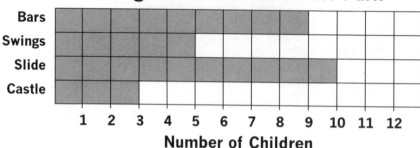

Mrs. Endo asked Kobe's class what they liked to do at the park. The graph shows how the children voted.

- **Kobe's favorite has an even number of votes.**
- **Branden's favorite has 4 more votes than the swings.**
- **Marcia's favorite has 1 less vote than the slide.**

What did Kobe, Branden, and Marcia vote for?

FIND OUT
What do you have to find out to solve the problem?

CHOOSE STRATEGIES
Ring the strategies you use.

SOLVE IT
Show your work.

LOOK BACK
Read the problem again. Check your work.

Garden Animals We Have Seen

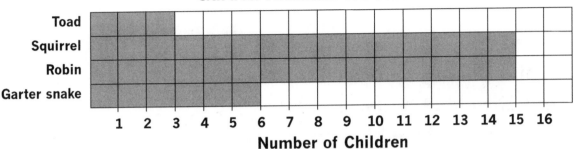

Number of Children

Mr. Brown asked his class which garden animals they had seen. The graph shows how many children saw each animal. Find out which animals fit these clues:

1. The fewest children saw this animal.

2. The same number of children that saw a squirrel saw this animal.

3. Three more children saw this animal than saw the toad.

Match each clue to a different animal.

FIND OUT

What do you have to find out to solve the problem?

CHOOSE STRATEGIES

Ring the strategies you use.

SOLVE IT

Show your work.

LOOK BACK

Read the problem again. Check your work.

Name _____

Jaime lives in the jungle. He loves to watch the birds. Today he sees 8 more yellow birds than blue birds. He sees the same number of blue birds as green birds. He sees 5 green birds. How many yellow birds does Jaime see today?

FIND OUT

What do you have to find out to solve the problem?

CHOOSE STRATEGIES

Ring the strategies you use.

SOLVE IT

Show your work.

LOOK BACK

Read the problem again. Check your work.

Name _____

In the Clowntown Circus there
are 4 more clowns than tumblers.
There are 7 more tumblers than
jugglers. There is 1 juggler in
the circus. How many clowns are
in the Clowntown Circus?

FIND OUT
What do you have to find out to solve the problem?

CHOOSE STRATEGIES
Ring the strategies you use.

SOLVE IT
Show your work.

LOOK BACK
Read the problem again. Check your work.

Name _____

Alec lives on a farm. He has 10 more chickens than rabbits. He has 3 fewer rabbits than ducks. He has 10 ducks. How many chickens does Alec have?

FIND OUT

What do you have to find out to solve the problem?

CHOOSE STRATEGIES

Ring the strategies you use.

SOLVE IT

Show your work.

LOOK BACK

Read the problem again. Check your work.

Name _____

Sandy and Petra went to see the movie
Miss Marble's Moose. It took them
1 hour to get to the movie. They were
at the movie for 2 hours. It took them
1 hour to get home. They got home
at 6 o'clock. What time did Sandy and
Petra leave for the movie?

FIND OUT
What do you have to find out to solve the problem?

CHOOSE STRATEGIES
Ring the strategies you use.

SOLVE IT
Show your work.

LOOK BACK
Read the problem again. Check your work.

Name _____

One hot day Gavin and Terry went swimming for 2 hours. Then they played ball for 1 hour. Next they went swimming for another 2 hours. Then it was 5:00. What time did Gavin and Terry first start swimming?

FIND OUT

What do you have to find out to solve the problem?

CHOOSE STRATEGIES

Ring the strategies you use.

SOLVE IT

Show your work.

LOOK BACK

Read the problem again. Check your work.

Permission is given to instructors to reproduce this page for classroom use with *Problem Solver II*. Copyright ©2004 Wright Group/McGraw-Hill

Name _____

Arlo and his family took a trip on Sunday.
They left home and drove for 1 hour. They
spent 2 hours at the dinosaur museum.
Then they spent 1 hour at a big park. They
drove home for 1 hour. They got home at 3:00.
When did Arlo and his family leave home?

FIND OUT

What do you have to find out to solve the problem?

CHOOSE STRATEGIES

Ring the strategies you use.

SOLVE IT

Show your work.

LOOK BACK

Read the problem again. Check your work.

81

Name _____

Lucas is fishing from a dock. In the
first hour he catches 4 fish. In the
second hour he catches 7 fish. In
the third hour he catches 10 fish.
In the fourth hour he catches 13 fish.
He keeps catching fish in the same
way. How many fish will Lucas catch
in the sixth hour?

FIND OUT
What do you have to find out to solve the problem?

CHOOSE STRATEGIES
Ring the strategies you use.

SOLVE IT
Show your work.

LOOK BACK
Read the problem again. Check your work.

Name _____

Little Red Riding Hood is worried. At 9:00 this morning she had 30 cakes in her basket. At 10:00 she had 25 cakes in her basket. At 11:00 she had 20 cakes in her basket. At 12:00 she had 15 cakes in her basket. The cakes keep disappearing in the same way. What time will it be when Little Red Riding Hood's basket is empty?

FIND OUT
What do you have to find out to solve the problem?

CHOOSE STRATEGIES
Ring the strategies you use.

SOLVE IT
Show your work.

LOOK BACK
Read the problem again. Check your work.

Name _____

On Monday 6 monkeys came to play by
the river. On Tuesday 12 monkeys came
to play. On Wednesday 18 monkeys
came to the river. The monkeys keep
coming in the same way. How many
monkeys will be at the river on Saturday?

FIND OUT

What do you have to find out to solve the problem?

CHOOSE STRATEGIES

Ring the strategies you use.

SOLVE IT

Show your work.

LOOK BACK

Read the problem again. Check your work.

Permission is given to instructors to reproduce this page for classroom use with *Problem Solver II*. Copyright ©2004 Wright Group/McGraw-Hill

84

Name _____

Tabby is making a puzzle about a mystery shape.
She draws these shapes:

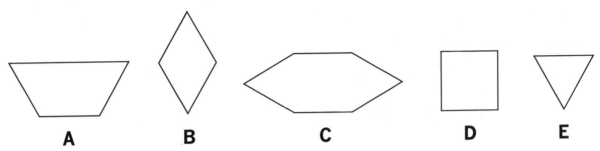

A   **B**   **C**   **D**   **E**

She gives these clues about her mystery shape:

- It is not a hexagon.
- It is not a rectangle.
- It is not a trapezoid.
- Its corners are not all the same size.

Which shape is Tabby's mystery shape?

FIND OUT

What do you have to find out to solve the problem?

CHOOSE STRATEGIES

Ring the strategies you use.

SOLVE IT

Show your work.

LOOK BACK

Read the problem again. Check your work.

Name _____

There is a Bean Bag Toss at the fair. If your bean bag hits the Lucky Shape, you win a prize! These clues tell about the Lucky Shape:

- It has an even number of sides.
- Its sides are all the same length.
- It does not have any square corners.

Which of these shapes is the Lucky Shape?

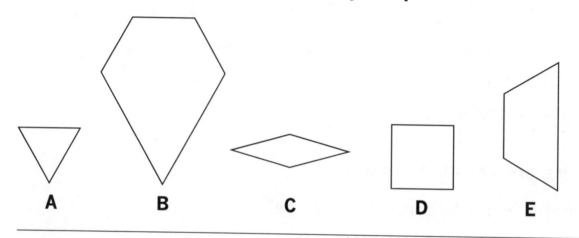

FIND OUT
What do you have to find out to solve the problem?

CHOOSE STRATEGIES
Ring the strategies you use.

SOLVE IT
Show your work.

LOOK BACK
Read the problem again. Check your work.

86

Name _____

Isabel wrote these clues about a mystery shape:

- **The shape has fewer sides than a hexagon.**
- **The corners of the shape are all the same size.**
- **The shape is a rectangle.**

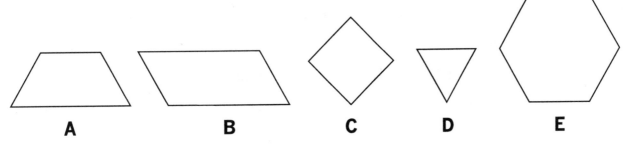

A   B   C   D   E

Which shape is Isabel's mystery shape?

_____

FIND OUT

What do you have to find out to solve the problem?

CHOOSE STRATEGIES

Ring the strategies you use.

SOLVE IT

Show your work.

LOOK BACK

Read the problem again. Check your work.

Name _____

A group of children from the Summer Day
Camp went to Funland Park. There were
15 children who went on the Pirate Ship ride.
There were 11 children who went on the
Haunted Cave ride. Six of the children went
on both rides. How many children in all went
to Funland from the Summer Day Camp?

FIND OUT

What do you have to find out to solve the problem?

CHOOSE STRATEGIES

Ring the strategies you use.

SOLVE IT

Show your work.

LOOK BACK

Read the problem again. Check your work.

Name _____

Cody talked to 14 first graders about their families. Each child had at least 1 sister, 1 brother, or both. Four of the children said they had both a sister AND a brother. There were 6 children who had ONLY a brother. How many children had ONLY a sister?

FIND OUT

What do you have to find out to solve the problem?

CHOOSE STRATEGIES

Ring the strategies you use.

SOLVE IT

Show your work.

LOOK BACK

Read the problem again. Check your work.

Name _____

There were 16 first graders at the carnival. They all played Air Hockey, Bean Guess, or both games. Two of the children played both Air Hockey AND Bean Guess. There were 8 children who played ONLY Bean Guess. How many children played ONLY Air Hockey?

FIND OUT
What do you have to find out to solve the problem?

CHOOSE STRATEGIES
Ring the strategies you use.

SOLVE IT
Show your work.

LOOK BACK
Read the problem again. Check your work.

Name _____

Tara made this shape with Pattern Block pieces:

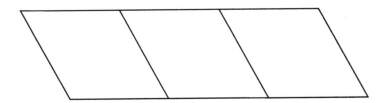

Then she said to Krista, "Can you solve my puzzle? Move 2 pieces and make a shape with 6 sides."

Can you solve Tara's problem?

FIND OUT
What do you have to find out to solve the problem?

CHOOSE STRATEGIES
Ring the strategies you use.

SOLVE IT
Show your work.

LOOK BACK
Read the problem again. Check your work.

91

Roberto gave this code puzzle to Adam.

I H 2 L   D H 4 L   B H 2 L   C H 4 L

Adam said, "I know what it means!"
Can you solve this code puzzle?

92

Debra made this shape with Pattern Blocks:

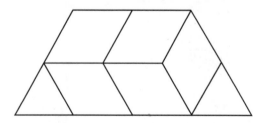

Then she said to Sofie, "Can you move only 1 piece
and make this into a parallelogram?"
Can you solve this puzzle?

Hint: You may find more than one answer.

93

At the Fall Harvest Fair,

 costs the same as

If 1  costs 15 cents, how much does 1  cost?

94

In Pico's Pet Shop,

1 Chewy Bone costs the same as 6 Munchy Sticks.

If 1 costs 10 cents, how much does 1  cost?

In the Swap Shop,

 costs the same as

If 1  costs 15 cents, how much does 1 ![hat] cost?

**Old King Cole is counting out his money.
He has 14 coins in all. He has 2 more quarters
than dimes. He has the same number of dimes
as nickels. How many nickels, dimes, and
quarters does Old King Cole have? How much
are his coins worth?**

97

Name _____

Chad shows Asia his calculator. This is what
Asia sees in the display:

| 36 |

Chad pushes a key. Now Asia sees this:

| 33 |

Chad pushes the key again. This time Asia sees:

| 30 |

Chad keeps pushing the key in the same way.
What will Asia see after Chad pushes the key
5 more times?

98

Name _____

At the Sunday movies, 60 children
in all saw *Ice Monsters*. There
were 45 children in all who saw
Jake, the Talking Cat. Twenty-five
of the children saw both movies.
How many children in all went to
these movies on Sunday?

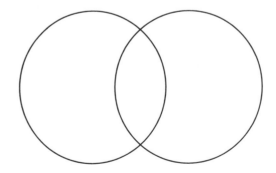

Our Favorite Pets

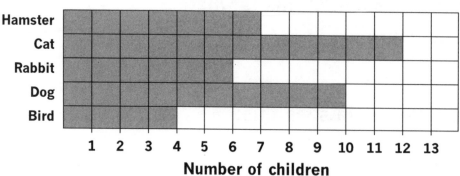

Two first-grade classes voted for their favorite pets.
The graph shows how many children voted for each pet.

- Jacob voted for the animal that got 6 fewer votes than the cat.
- Mona voted for the animal that got twice as many votes as the rabbit.
- Missy voted for the animal that got an odd number of votes.

Which pets did Jacob, Mona, and Missy vote for?

Permission is given to instructors to reproduce this page for classroom use with *Problem Solver II*.
Copyright ©2004 Wright Group/McGraw-Hill

The squirrels are eating Alison's peaches.
On Saturday there were 24 peaches on the tree.
On Sunday there were 20 peaches on the
tree. On Monday there were 16 peaches on the
tree. On Tuesday there were 12 peaches on
the tree. The peaches keep disappearing in the
same way. What day will the peaches be gone?

Permission is given to instructors to reproduce this page for classroom use with *Problem Solver II*.
Copyright ©2004 Wright Group/McGraw-Hill

Answer Key

Teaching Problem/Solution

Note: The solution process is included in each lesson for Teaching Problems 1–38.

---

1 Table A – 1 cat, table B – 4 cats, table C – 3 cats

2 First plate – 6 cookies, second plate – 2 cookies, third plate – 4 cookies

3 9 puppets

4 8 robots

5 5 nickels

6 9 cards

7 6 red fish, 3 green fish

8 8 nickels, 4 pennies

9 7 times

10 11 times

11 1 – green, 2 – blue, 3 – red, 4 – yellow

12 1 – red, 2 – blue, 3 – red, 4 – green

13 penny – penny, nickel – nickel, dime – dime, penny – nickel; or penny – penny, nickel – nickel, penny – dime, nickel – dime

14 Solutions include: yellow – yellow, green – green, blue – blue, red – red, yellow – green; or yellow – green, yellow – blue, yellow – red, green – green, red – blue

15

Pocket A	Pocket B
1	4
2	3
3	2
4	1

16

Bowl A	Bowl B
1	5
2	4
3	3
4	2
5	1

17 Sue is ahead by 1 space.

18 Dom is on step 9.

19 12 cents (2 pennies, 2 nickels)

20 21 cents (2 nickels, 1 penny, 1 dime)

21 Peanut butter

22 Dog food

23 Angela – banana, Shane – orange, Lorie – apple

24 Joey – Match It, Maria – Go Fish, Beth – Match It

25 10 red flowers

26 10 blue puzzle pieces

27 1 o'clock

28 3 o'clock

29 20 bones

30 Monday

31 The hexagon

32 The trapezoid

33 8 friends

34 15 children

35 Switch the penny at the bottom of the first column with the nickel at the top of the third column.

36 2 eyes, 2 ears, 1 nose, 1 mouth

37 15 cents

38 40 cents

Practice Problem/Solution

Note: The strategies shown for the Practice Problems are those which were used for solving the similar Teaching Problems. However, students' choice of strategy may vary.

39 First boat – 7 pirates, second boat – 5 pirates, third boat – 3 pirates

40 First bag – 1 dino, second bag – 7 dinos, third bag – 9 dinos

41 First pile – 11 pennies, second pile – 5 pennies, third pile – 2 pennies

42 7 hats

Number of Hats	Number of Frogs
1	4
2	8
3	12
4	16
5	20
6	24
7	28

43 6 cars

Number of Cars	Number of Children
1	5
2	10
3	15
4	20
5	25
6	30

44 6 times

Times Landed on Hippo	Number of Points
1	6
2	12
3	18
4	24
5	30
6	36

45 12 bears

Day	Number of Bears
1st	2
2nd	4
3rd	6
4th	8
5th	10
6th	12

46 15 flowers

Day	Number of Flowers
1st	3
2nd	6
3rd	9
4th	12
5th	15

47 15 stories

Day	Number of Stories
Sunday	3
Monday	5
Tuesday	7
Wednesday	9
Thursday	11
Friday	13
Saturday	15

48 10 purple monsters, 5 orange monsters

49 11 pennies, 5 dimes

50 3 red T-shirts, 3 blue T-shirts, 5 pink T-shirts

51 18 lizards

Day	Number of Lizards
1st	3
2nd	6
3rd	9
4th	12
5th	15
6th	18

52 13 bricks

Day	Number of Bricks
Monday	3
Tuesday	5
Wednesday	7
Thursday	9
Friday	11
Saturday	13

53 5 ears of corn

Day	Ears of Corn
Monday	20
Tuesday	17
Wednesday	14
Thursday	11
Friday	8
Saturday	5

54 1 – white, 2 – red, 3 – yellow, 4 – blue, 5 – white

55 1 – blue, 2 – yellow, 3 – red, 4 – red, 5 – green

56 1 – yellow, 2 – red, 3 – yellow, 4 – blue, 5 – red, 6 – green

57 Penny – penny, nickel – nickel, dime – dime, nickel – penny, nickel – dime, dime – penny

58 Solutions include: yellow – yellow, green – green, red – red, blue – blue, blue – red, yellow – green; or yellow – green, yellow – red, yellow – blue, green – red, green – blue, red – blue

59 Solutions include: blue – blue, yellow – green, yellow – red, green – green, red – green; or blue – yellow, blue – green, green – green, yellow – green, red – red

60

Pot A	Pot B
1	6
2	5
3	4
4	3
5	2
6	1

61

Bank A	Bank B
1	7
2	6
3	5
4	4
5	3
6	2
7	1

62

Boat A	Boat B
1	8
2	7
3	6
4	5
5	4
6	3
7	2
8	1

63 Lisette is ahead; her marker is on space 15, and Jodi's marker is on space 13.

1	2	3	4	5	6	7	8	9	10	11	12	13	14	15	16
X			X	X								X		X	

Jodi   Lisette

64 11th step

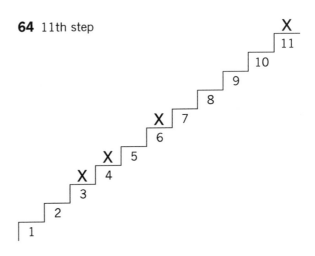

65 Locker 7

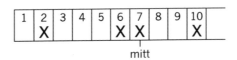

1	2	3	4	5	6	7	8	9	10
	X				X	X			X

mitt

66 28 cents (3 pennies, 1 nickel, 2 dimes)

67 36 cents (3 nickels, 1 penny, 2 dimes)

68 40 cents (1 quarter, 1 nickel, 1 dime)

69 Billy misses the berry pie.

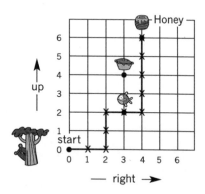

70 Franny eats 5 flies.

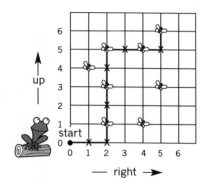

71 Any solutions that include 7 hops to the right and 5 hops forward are correct, except the same path as that taken by the mother kangaroo.

72 Melody – merry-go-round, Julie – roller coaster, Andre – boat

73 Kobe – slide, Branden – bars, Marcia – bars

74 Clue 1 – toad, clue 2 – robin, clue 3 – garter snake

75 13 yellow birds

76 12 clowns

77 17 chickens

78 2 o'clock

79 12 o'clock

80 10 o'clock

81 19 fish

Hour	Number of Fish
1st	4
2nd	7
3rd	10
4th	13
5th	16
6th	19

82 3 o'clock

Hour	Number of Cakes
9 o'clock	30
10 o'clock	25
11 o'clock	20
12 o'clock	15
1 o'clock	10
2 o'clock	5
3 o'clock	0

83 36 monkeys

Day	Number of Monkeys
Monday	6
Tuesday	12
Wednesday	18
Thursday	24
Friday	30
Saturday	36

84 Shape B, the parallelogram. The first clue eliminates shape C; the second clue eliminates shape D; the third clue eliminates shape A; the fourth clue eliminates shape E.

85 Shape C, the parallelogram. The first clue eliminates the triangle and pentagon; the second clue eliminates the trapezoid; the third clue eliminates the square.

86 Shape C, the square. The first clue eliminates the hexagon; the second clue eliminates the trapezoid and parallelogram; the third clue eliminates the triangle.

87 20 children

Pirate Ship   **Haunted Cave**

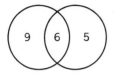

88 4 children had only a sister.

Sister   **Brother**

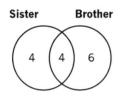

89 6 children played only Air Hockey.

Air Hockey   **Bean Guess**

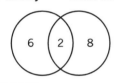

90

91 I have 2 legs. Dogs have 4 legs. Birds have 2 legs. Cats have 4 legs. (Other answers are possible.)

92

Or, students might move the left triangle up to the top left corner.

93 30 cents. If 1 apple costs 15 cents, then 2 apples cost 30 cents. If 1 ear of corn costs the same as 2 apples, then the corn costs 30 cents.

94 60 cents. If 1 Munchy Stick costs 10 cents, then 6 Munchy Sticks cost 60 cents. If 1 Chewy Bone costs the same as 6 Munchy Sticks, then 1 Chewy Bone costs 60 cents.

95 60 cents. If 1 magnifying glass costs 15 cents, then 4 of them cost 60 cents. If 1 detective cap costs the same as 4 magnifying glasses, then the cap costs 60 cents.

96 4 nickels, 4 dimes, 6 quarters; $2.10.

97 Asia will see ⟦ 15 ⟧ in the calculator display.

Number of Pushes	Number in Display
1	36
2	33
3	30
4	27
5	24
6	21
7	18
8	15

98 80 children

Ice Monsters   **Jake**

35 (25) 20

99 Jacob – rabbit, Mona – cat, Missy – hamster

100 Friday

Day	Number of Peaches
Saturday	24
Sunday	20
Monday	16
Tuesday	12
Wednesday	8
Thursday	4
Friday	0

Resources

Books

Bransford, John D., Ann L. Brown, and Rodney R. Cocking, eds. *How People Learn: Brain, Mind, Experience, and School.* Washington D.C.: National Academy Press, 2000.

Burns, Marilyn. *About Teaching Mathematics: A K–8 Resource.* 2nd ed. Sausalito, CA: Math Solutions Publications, 2000.

National Council of Teachers of Mathematics. *Principles and Standards for School Mathematics.* Reston, VA: National Council of Teachers of Mathematics, 2000.

Polya, George. *How to Solve It: A New Aspect of Mathematical Method.* 2nd ed. Princeton, NJ: Princeton University Press, 1957.

Van De Walle, John A. *Elementary and Middle School Mathematics: Teaching Developmentally.* 4th ed. Addison Wesley Longman, Inc., 2001.

Web Sites

Web sites tend to change, so we hope that these sites remain useful and in operation. We believe that they are maintained by outstanding organizations.

www.aimsedu.org
Good teacher resources and a Puzzle Corner with a variety of problems for students to solve.

www.bigchalk.com
Math resources for parents, teachers, and students.

www.coolmath4kids.com
A visually inviting site with activities for all ages.

www.dupagechildrensmuseum.org/aunty
"Aunty Math" presents math challenges for grades K–5. Includes information about problem-solving strategies for parents and teachers, along with suggestions for modifying problems to make them easier or more challenging.

www.eduplace.com/math/brain
Weekly brainteasers for students in grades 3–8. Site maintained by Houghton Mifflin Company.

www.mathforum.org/elempow
Problems of the week for elementary students, including a bank of problems with solutions, indexed by math topic. Site operated by Drexel University.

www.math.rice.edu/~lanius/Lessons
Interactive activities with inviting graphics, offering math challenges for variety of ages.

www.nrich.maths.org
Wonderful problem archive with hundreds of excellent problems clearly marked for appropriate ages. Site operated by University of Cambridge.

www.olemiss.edu/mathed/brain
Problem of the week brainteasers for elementary students. Site operated by University of Mississippi.

www.techlearning.com
Resources on educational technology for teachers, with Curriculum Resource links to recommended math sites.